John J. Masterson
D-211 Wells Hall.
Michigan State Univ
E. Lansing MI 48824

A Transition to Advanced Mathematics

A Transition to Advanced Mathematics

Douglas Smith
Central Michigan University
Maurice Eggen
Trinity University
Richard St. Andre
Central Michigan University

Brooks/Cole Publishing Company
Monterey, California

Brooks/Cole Publishing Company
A Division of Wadsworth, Inc.

Printed in the United States of America

10 9 8 7 6 5 4 3

Library of Congress Cataloging in Publication Data

Smith, Douglas, [date]
Transition to advanced mathematics.

Includes index.
1. Mathematics—1961- . I. Eggen, Maurice,
[date]. II. St. Andre, Richard, [date].
III. Title.
QA37.2.S575 1983 510 82-20737

ISBN 0-534-01249-3

Subject Editor: Craig Barth
Manuscript Editor: Kirk Sargent
Production Editor: Joan Marsh
Interior and Cover Design: Vicki Van Deventer
Illustrations: John Foster
Typesetting: Omegatype Typography, Inc.

To Karen, Karen, and Karen

Preface

"I understand mathematics but I just can't do proofs."

Our experience has led us to believe that the remark above, though contradictory, expresses the frustration many students feel as they pass from beginning calculus to a more rigorous level of mathematics. This book developed from a series of lecture notes for a course at Central Michigan University that was designed to address this lament. The text is intended to bridge the gap between calculus and advanced courses in at least three ways. First, it provides a firm foundation in the major ideas needed for continued work. Second, it guides students to think and to express themselves mathematically—to analyze a situation, extract pertinent facts, and draw appropriate conclusions. Finally, we present introductions to modern algebra and analysis of sufficient depth to capture some of their spirit and characteristics.

We begin in Chapter 1 with a study of the logic required by mathematical arguments, discussing not formal logic but rather the standard methods of mathematical proof and their validity. Methods of proof are examined in detail, and examples of each method are analyzed carefully. Denials are given special attention, particularly those involving quantifiers. Techniques of proof given in this chapter are used and referred to later in the text. Although the chapter was written with the idea that it may be assigned as out-of-class reading, we find that most students benefit from a thorough study of logic.

Much of the material in Chapters 2, 3, and 4 on sets, relations, and functions, will be familiar to the student. Thus, the emphasis is on enhancing the student's ability to write and understand proofs. The pace is deliberate. The rigorous approach requires the student to deal precisely with these concepts.

Chapters 5, 6, and 7 make use of the skills and techniques the student has acquired in Chapters 1 through 4. These last three chapters are a cut above the earlier chapters in terms of level and rigor. *Chapters 1 through 4 and* ***any one of*** *Chapters 5, 6, or 7 provide sufficient material for a one-semester course.* An alternative is to choose among topics by selecting, for example, the first two sections of Chapter 5, the first three sections of Chapter 6, and the first two sections of Chapter 7.

Chapter 5 begins the study of cardinality by examining the properties of finite and infinite sets and establishing countability or uncountability for the familiar number systems. The emphasis is on a working knowledge of cardinality—particularly countable sets, the ordering of cardinal numbers, and applications of the Cantor-Schröeder-Bernstein Theorem. We include a brief discussion of the Axiom of Choice and relate it to the comparability of cardinals.

Chapter 6, which introduces modern algebra, concentrates on the concept of a group and culminates in the Fundamental Theorem of Group Homomorphisms. The idea of an operation preserving map is introduced early and developed throughout the section. Permutation groups, cyclic groups, and modular arithmetic are among the examples of groups presented.

Chapter 7 begins with a description of the real numbers as a complete ordered field. We continue with the Heine-Borel Theorem, the Bolzano-Weierstrass Theorem, and the Bounded Monotone Sequence Theorem (each for the real number system), and then return to the concept of completeness.

Exercises marked with a solid star ★ have complete answers at the back of the text. Open stars ☆ indicate that a hint or a partial answer is provided. "Proofs to Grade" are a special feature of most of the exercise sets. We present a list of claims with alleged proofs, and the student is asked to assign a letter grade to each "proof" and to justify the grade assigned. Spurious proofs are usually built around a single type of error, which may involve a mistake in logic, a common misunderstanding of the concepts being studied, or an incorrect symbolic argument. Correct proofs may be straightforward, or they may present novel or alternate approaches. We have found these exercises valuable because they reemphasize the theorems and counterexamples in the text and also provide the student with an experience similar to grading papers. Thus, the student becomes aware of the variety of possible errors and develops the ability to read proofs critically.

In summary, our main goals in this text are to improve the student's ability to think and write in a mature mathematical fashion and to provide a solid understanding of the material most useful for advanced courses. Student readers, take comfort from the fact that we do not aim to turn you into theorem-proving wizards. Few of you will become research mathematicians. Nevertheless, in almost any mathematically related work you may do, the kind of reasoning you need to be able to do is the same reasoning you use in proving theorems. You must first understand exactly what you want to prove (verify, show, or explain), and you must be familiar with the logical steps that allow you to get from the hypothesis to the conclusion. Moreover, a proof is the ultimate test of your understanding of the subject matter and of mathematical reasoning.

We are grateful to the many students who endured earlier versions of the manuscript and gleefully pointed out misprints. We acknowledge also the helpful comments of Edwin H. Kaufman, Melvin Nyman, Mary R. Wardrop, and especially Douglas W. Nance, who saw the need for a course of this kind at CMU and did a superb job of reviewing the manuscript.

We thank our reviewers: William Ballard of the University of Montana, Sherralyn Craven of Central Missouri State University, Robert Dean of Stephen F. Austin State University, Harvey Elder of Murray State University, Joseph H. Oppenheim of San Francisco State University, Joseph Teeters of the University of Wisconsin, Dale Schoenefeld of the University of Tulsa, Kenneth Slonnegar of State University of New York at Fredonia, and Douglas Smith of University of the Pacific.

We are grateful to Craig Barth, Mathematics Editor at Brooks/Cole, for his encouragement and faith in the project and to Joan Marsh and the editorial staff at Brooks/Cole for their excellent work. And we wish to thank Karen St. Andre for her superb and expeditious typing of the manuscript.

Douglas Smith
Maurice Eggen
Richard St. Andre

Contents

1 **Logic and Proofs 1**

1.1 Propositions and Connectives 1
1.2 Conditionals and Biconditionals 5
1.3 Quantifiers 11
1.4 Mathematical Proofs 15
1.5 Proofs Involving Quantifiers 24

2 **Set Theory 30**

2.1 Basic Notions of Set Theory 30
2.2 Set Operations 36
2.3 Extended Set Operations and Indexed Families of Sets 41
2.4 Induction 47

3 **Relations 54**

3.1 Cartesian Products and Relations 54
3.2 Equivalence Relations 62
3.3 Partitions 66

4 **Functions 70**

4.1 Functions as Relations 70
4.2 Constructions of Functions 75
4.3 Onto Functions; One-to-One Functions 81
4.4 Induced Set Functions 87

*5 **Cardinality 92**

5.1 Equivalent Sets; Finite Sets 92
5.2 Infinite Sets 96

*Optional Material

5.3 The Ordering of Cardinal Numbers 103
5.4 Comparability of Cardinals and the Axiom of Choice 108
5.5 Countable Sets 111

***6 Concepts of Algebra 114**

6.1 Algebraic Structures 114
6.2 Groups 120
6.3 Examples of Groups 124
6.4 Subgroups 128
6.5 Cosets and Lagrange's Theorem 133
6.6 Quotient Groups 135
6.7 Isomorphism; The Fundamental Theorem of Group Homomorphisms 139

***7 Real Analysis 143**

7.1 Field Properties of the Real Numbers 143
7.2 The Heine-Borel Theorem 147
7.3 The Bolzano-Weierstrass Theorem 153
7.4 The Bounded Monotone Sequence Theorem 156
7.5 Equivalents of Completeness 160

Answers to Selected Problems 162

Index 175

List of Symbols 178

A Transition to Advanced Mathematics

Logic and Proofs 1

One difference that may be observed between mathematicians and other scientists is in the general approach of each to work they undertake. The natural or social scientist frequently uses an empirical approach and proceeds from the observation of particular cases to a general theory by inductive reasoning. The validity of the theory is then tested on other cases, and if the results are incompatible with theoretical expectations, the scientist must reject or modify his theory. A mathematician, who characteristically employs deductive reasoning, develops a theory step by step, in a logically valid fashion, to insure his conclusions are correct. Then, if the theory predicts results incompatible with reality, the fault will lie not in the theory but with the inapplicability of the theory to that portion of reality.

The goal of this chapter is to provide a working knowledge of the basics of logic and the idea of proof, which are fundamental to deductive reasoning.

SECTION 1.1. PROPOSITIONS AND CONNECTIVES

A **proposition** is a sentence that is either true or false. Some examples of propositions are "$\sqrt{2}$ is irrational," "The world will end in 4821," and "$1 + 1 = 5$." "What did you say?" and "This sentence is false"[†] are not propositions.

The propositions listed above are simple, in the sense that they do not have any other propositions as components. Simple propositions can be put together using logical connectives to form compound propositions.

Definitions. Given propositions P and Q,

- The **conjunction of *P* and *Q*,** denoted $P \wedge Q$, is the proposition "P and Q." $P \wedge Q$ is true exactly when both P and Q are true.
- The **disjunction of *P* and *Q*,** denoted $P \vee Q$, is the proposition "P or Q." $P \vee Q$ is true exactly when at least one of P or Q is true.
- The **negation of *P*,** denoted $\sim P$, is the proposition "not P." $\sim P$ is true exactly when P is false.

[†] This is an example of a sentence that is neither true nor false. The study of paradoxes such as this has played a key role in the development of modern mathematical logic.

If P is "$1 < 3$" and Q is "7 is odd," then $P \wedge Q$ is "$1 < 3$ and 7 is odd"; $P \vee Q$ is "$1 < 3$ or 7 is odd" and $\sim Q$ is "It is not the case that 7 is odd," which of course is to say "7 is even." Since both P and Q are true, $P \wedge Q$ and $P \vee Q$ are true, while $\sim Q$ is false.

All of the following are true propositions:

"It is not the case that $\sqrt{10} > 4$."
"$\sqrt{2} < \sqrt{3}$ or chickens have lips."
"Venus is smaller than the earth or $1 + 4 = 5$."
"$6 < 7$ and $7 < 8$."

All of the following are false:

"1955 was a bad year for wine and π is rational."
"It is not the case that 10 is divisible by 2."
"$2^4 = 16$ and a quart is larger than a liter."

Other connectives commonly used in English are "but," "while," "although," and so on. Each of these named would normally be translated symbolically with the conjunction connective. A variant of the connective "or" is discussed in exercise 7.

The truth values of a compound proposition are readily obtained by exhibiting all possible combinations of the truth values for its components in a truth table. Since each connective $\wedge$ and $\vee$ involves two components their truth tables must list the four possible combinations of the truth values of those components. When we use T for true and F for false, the truth tables for $P \wedge Q$ and $P \vee Q$ are

P	Q	$P \wedge Q$
T	T	T
F	T	F
T	F	F
F	F	F

P	Q	$P \vee Q$
T	T	T
F	T	T
T	F	T
F	F	F

Since the value of $\sim P$ depends only on the two possible values for P, its truth table is

P	$\sim P$
T	F
F	T

Frequently you will encounter compound propositions with more than two simple components. The proposition $(\sim Q \vee P) \wedge (R \vee S)$ has four simple components; it follows that there are $2^4 = 16$ possible combinations of values for P, Q, R, S. The two main components are $\sim Q \vee P$ and $R \vee S$. One way to make the table for $(\sim Q \vee P) \wedge (R \vee S)$ is to make tables first for both of these components and then combine those values by using the truth table for $\wedge$.

P	Q	R	S	$\sim Q$	$\sim Q \vee P$	$R \vee S$	$(\sim Q \vee P) \wedge (R \vee S)$
T	T	T	T	F	T	T	T
F	T	T	T	F	F	T	F
T	F	T	T	T	T	T	T
F	F	T	T	T	T	T	T
T	T	F	T	F	T	T	T
F	T	F	T	F	F	T	F
T	F	F	T	T	T	T	T
F	F	F	T	T	T	T	T
T	T	T	F	F	T	T	T
F	T	T	F	F	F	T	F
T	F	T	F	T	T	T	T
F	F	T	F	T	T	T	T
T	T	F	F	F	T	F	F
F	T	F	F	F	F	F	F
T	F	F	F	T	T	F	F
F	F	F	F	T	T	F	F

Definition. Two propositions P and Q are **equivalent** if and only if they have the same truth tables.

For example, the propositions $P \vee (Q \wedge P)$ and P are equivalent. To show this we examine their truth tables.

P	Q	$Q \wedge P$	$P \vee (Q \wedge P)$
T	T	T	T
F	T	F	F
T	F	F	T
F	F	F	F

Since the P column and the $P \vee (Q \wedge P)$ column are identical, the propositions are equivalent.

Any proposition P is equivalent to itself. Also the propositions P and $\sim(\sim P)$ are equivalent. Their tables are

P	$\sim P$	$\sim(\sim P)$
T	F	T
F	T	F

Definition. A **denial** of a proposition S is any proposition equivalent to $\sim S$.

The proposition "$\sqrt{2}$ is rational," has the negation "It is not the case that $\sqrt{2}$ is rational," or "$\sqrt{2}$ is not rational." Some denials are "$\sqrt{2}$ is irrational," "$\sqrt{2}$ is not a repeating decimal," and "$\sqrt{2}$ cannot be written as the quotient of two integers."

It can be seen from the example above that the negation $\sim P$ is a denial of the proposition P but a denial need not be the negation. The ability to rewrite the negation of a proposition into a useful denial will be very important for writing indirect proofs. (See section 1.4.)

In order to avoid writing large numbers of parentheses, we use the rule that, first, $\sim$ applies to the smallest proposition following it, then $\wedge$ connects the smallest propositions surrounding it, and finally $\vee$ connects the smallest propositions surrounding it. Thus $\sim P \vee Q$ is an abbreviation for $(\sim P) \vee Q$, rather than $\sim(P \vee Q)$. The proposition $P \wedge \sim Q \vee R$ abbreviates $(P \wedge (\sim Q)) \vee R$.

Exercises 1.1

1. Make truth tables for each of the following propositions:
   - ★ (a) $P \wedge \sim P$ (b) $P \vee \sim P$
   - ★ (c) $P \wedge (Q \vee R)$ (d) $(P \wedge Q) \vee (P \wedge R)$
   - ★ (e) $P \wedge \sim Q$ (f) $P \wedge (Q \vee \sim Q)$
   - ★ (g) $(P \wedge Q) \vee \sim Q$ (h) $\sim(P \wedge Q)$
   - (i) $(P \vee \sim Q) \wedge R$ (j) $\sim P \wedge \sim Q$
   - (k) $P \wedge P$ (l) $(P \wedge Q) \vee (R \wedge \sim S)$
2. Which of the following pairs of propositions are equivalent?
   - ★ (a) $P \wedge P$, P (b) $P \vee P$, P
   - ★ (c) $P \wedge Q$, $Q \wedge P$ (d) $P \vee Q$, $Q \vee \sim P$
   - ★ (e) $(P \wedge Q) \wedge R$, $P \wedge (Q \wedge R)$ (f) $\sim(P \wedge Q)$, $\sim P \wedge \sim Q$
   - ★ (g) $\sim P \wedge \sim Q$, $\sim(P \wedge \sim Q)$ (h) $(P \vee Q) \vee R$, $P \vee (Q \vee R)$
   - ★ (i) $(P \wedge Q) \vee R$, $P \wedge (Q \vee R)$ (j) $\sim(P \vee Q)$, $\sim P \wedge \sim Q$
   - ★ (k) $\sim(P \wedge Q)$, $\sim P \vee \sim Q$ (l) $(P \wedge Q) \vee R$, $P \vee (Q \wedge R)$
   - ★ (m) $P \wedge (Q \vee R)$, $(P \wedge Q) \vee (P \wedge R)$ (n) $\sim P \vee \sim Q$, $\sim(P \vee \sim Q)$
3. If P, Q, and R are true while S and T are false, which of the following are true?
   - ★ (a) $Q \wedge (R \wedge S)$ (b) $Q \vee (R \wedge S)$
   - ★ (c) $(P \vee Q) \wedge (R \vee S)$ (d) $(\sim P \vee \sim Q) \vee (\sim R \vee \sim S)$
   - ★ (e) $\sim P \vee (Q \wedge \sim Q)$ (f) $\sim P \vee \sim Q$
   - ★ (g) $(\sim Q \vee S) \wedge (Q \vee S)$ (h) $(S \wedge R) \vee (S \wedge T)$
   - ★ (i) $(P \vee S) \wedge (P \vee T)$ (j) $(\sim T \wedge P) \vee (T \wedge P)$
   - ★ (k) $\sim P \wedge (Q \vee \sim Q)$ (l) $\sim R \wedge \sim S$
4. Give a useful denial of each:
   - ★ (a) x is a positive integer.
   - (b) She will marry Heckle or Jeckle.
   - ★ (c) $5 \geq 3$.
   - (d) 641371 is a composite integer.
   - ★ (e) Roses are red and violets are blue.

(f) $x < y$ or $m^2 < 1$.

★ (g) T is not green or T is yellow.

(h) She will choose yogurt but will not choose ice cream.

5. If P, Q, and R are propositions, and P is equivalent to Q, and Q is equivalent to R, prove that:

★ (a) Q is equivalent to P.

(b) P is equivalent to R.

(c) $\sim Q$ is equivalent to $\sim P$.

(d) $P \wedge Q$ is equivalent to $Q \wedge R$.

(e) $P \vee Q$ is equivalent to $Q \vee R$.

6. Let P be the sentence "Q is true" and Q be the sentence "P is false." Is P a proposition? Explain.

7. The word *or* is used in two different ways in English. We have presented the truth table for $\vee$, the **inclusive or** whose meaning is "one or the other or both." The **exclusive or** meaning "one or the other but not both," denoted $\textcircled{\vee}$, has its uses in English, as in "She will marry Heckle or she will marry Jeckle." The inclusive or is much more useful in mathematics and is the accepted meaning unless there is a statement to the contrary.

★ (a) Make a truth table for the "exclusive or" connective, $\textcircled{\vee}$.

(b) Show that $A \textcircled{\vee} B$ is equivalent to $(A \vee B) \wedge \sim(A \wedge B)$.

SECTION 1.2. CONDITIONALS AND BICONDITIONALS

It is fair to say that the most important kind of proposition is a sentence of the form "If P, then Q." Examples include "If I had the money, then I would buy a new car," "If two lines in a plane do not intersect, then they are parallel," and "If f is differentiable at x_0 and x_0 is a relative minimum for f, then $f'(x_0) = 0$."

Definition. Given propositions P and Q, the **conditional sentence** $P \Rightarrow Q$ (read "P implies Q") is the proposition "If P, then Q." The proposition P is the **antecedent** and Q is the **consequent.** The conditional sentence $P \Rightarrow Q$ is true whenever the antecedent is false or the consequent is true.

The truth table for $P \Rightarrow Q$ is

P	Q	$P \Rightarrow Q$
T	T	T
F	T	T
T	F	F
F	F	T

This table gives $P \Rightarrow Q$ the value F only when P is true and Q is false, and thus it agrees with the meaning of "if . . . , then . . ." in promises. For example, the person who promises, "If Lincoln was the second U.S. President, I'll give you a dollar" would not be called a liar for failing to give you a dollar. In fact, he could give you a dollar and still not be a liar. In both cases we say the statement is true because the antecedent is false.

The situation in mathematics is similar. We all agree that "If x is an odd integer, then $x + 1$ is even" is a true proposition. It would be hopeless to protest that in the case where x is 6, then $x + 1$ is 7, which is not even, for after all, we only claim that *if* x is odd, *then* $x + 1$ is even.

One curious consequence of the use of the truth table for $P \Rightarrow Q$ is that conditional sentences may be true even when there is no connection between the antecedent and the consequent.

For example, all of the following are true:

- $\text{Sin } 30° = \frac{1}{2} \Rightarrow 1 + 1 = 2$.
- Mars has ten moons $\Rightarrow 1 + 1 = 2$.
- Mars has ten moons $\Rightarrow$ Paul Revere made plastic spoons.

And the following are false:

- $1 + 2 = 3 \Rightarrow 1 < 0$.
- Ducks have webbed feet $\Rightarrow$ The U.S. National Debt is $915.45.

Two propositions closely related to the conditional sentence $P \Rightarrow Q$ are its converse and its contrapositive.

Definition. For propositions P and Q, the **converse of $P \Rightarrow Q$** is $Q \Rightarrow P$ and the **contrapositive of $P \Rightarrow Q$** is $\sim Q \Rightarrow \sim P$.

For the conditional sentence "If a function f is differentiable at x_0, then f is continuous at x_0," its converse is "If f is continuous at x_0, then f is differentiable at x_0," while the contrapositive is "If f is not continuous at x_0, then f is not differentiable at x_0." You should know from calculus which of these are true statements.

If P is the proposition "It is raining here" and Q is "It is cloudy overhead," then $P \Rightarrow Q$ is true. Its contrapositive is "If it is not cloudy overhead, then it is not raining here," which is also true. However, the converse "If it is cloudy overhead, then it is raining here" is not a true proposition. The relationships between a conditional sentence and its contrapositive and converse are given in the following theorem.

Theorem 1.1. For any propositions P and Q,

(a) $P \Rightarrow Q$ is equivalent to $\sim Q \Rightarrow \sim P$.
(b) $P \Rightarrow Q$ is *not* equivalent to $Q \Rightarrow P$.

Proof. A proof requires examining the truth tables:

P	Q	$P \Rightarrow Q$	$\sim Q$	$\sim P$	$\sim Q \Rightarrow \sim P$	$Q \Rightarrow P$
T	T	T	F	F	T	T
F	T	T	F	T	T	F
T	F	F	T	F	F	T
F	F	T	T	T	T	T

Comparing the third and sixth columns we conclude $P \Rightarrow Q$ is equivalent to $\sim Q \Rightarrow \sim P$. Comparing the third and seventh columns, we see they differ in the second and third line. Thus $P \Rightarrow Q$ and $Q \Rightarrow P$ are not equivalent. ■

The equivalence of a conditional sentence and its contrapositive will provide an important proof technique developed in section 1.4. (See page 20.) The truthfulness or falsity of a conditional sentence has no influence on the truth value of the converse. The converse cannot be used to prove a conditional sentence.

Closely related to the conditional sentence is the biconditional sentence $P \Leftrightarrow Q$. The double arrow $\Leftrightarrow$ reminds one of both $\Leftarrow$ and $\Rightarrow$, and this is not an accident. We shall soon see that $P \Leftrightarrow Q$ is equivalent to $(P \Rightarrow Q) \wedge (Q \Rightarrow P)$.

Definition. For propositions P and Q, the **biconditional sentence $P \Leftrightarrow Q$** is the proposition "P if and only if Q." The sentence $P \Leftrightarrow Q$ is true exactly when P and Q have the same truth values.

The truth table for $P \Leftrightarrow Q$ is

P	Q	$P \Leftrightarrow Q$
T	T	T
F	T	F
T	F	F
F	F	T

As a form of shorthand the words "if and only if" are frequently abbreviated to "iff" in mathematics. The statements "A rectangle is a square iff the rectangle's diagonals are perpendicular" and "$1 + 7 = 6$ iff $\sqrt{2} + \sqrt{3} = \sqrt{5}$" are both true biconditional sentences, while "Lake Erie is in Peru iff π is an irrational number" is a false biconditional sentence.

Because the biconditional sentence $P \Leftrightarrow Q$ has the value T exactly when the values of P and Q are the same, the biconditional connective can be used to test whether P and Q are equivalent: $P \Leftrightarrow Q$ will be true precisely when P is equivalent to Q. For example, the sentence P given by "3 and -3 have the same absolute value" and the sentence Q given by "3 and -3 have equal squares" are equivalent since $P \Leftrightarrow Q$ is true. Any properly stated definition is an example of a biconditional sentence.

One key to success in mathematics is the ability to replace a statement by a more useful or enlightening equivalent one. This is precisely what you do to "solve" $x^2 - 7x = -12$.

$$\begin{aligned} x^2 - 7x &= -12 & \Leftrightarrow \\ x^2 - 7x + 12 &= 0 & \Leftrightarrow \\ (x-3)(x-4) &= 0 & \Leftrightarrow \\ x - 3 = 0 \text{ or } x - 4 &= 0 & \Leftrightarrow \\ x = 3 \text{ or } x &= 4. \end{aligned}$$

Each statement is simply an equivalent of its predecessor but is more illuminating as to the solution. The ability to write equivalents is crucial in writing proofs. The next theorem contains seven important equivalences.

Theorem 1.2. For propositions P and Q,

(a) $P \Leftrightarrow Q$ is equivalent to $(P \Rightarrow Q) \wedge (Q \Rightarrow P)$.
(b) $\sim(P \wedge Q)$ is equivalent to $\sim P \vee \sim Q$.
(c) $\sim(P \vee Q)$ is equivalent to $\sim P \wedge \sim Q$.
(d) $\sim(P \Rightarrow Q)$ is equivalent to $P \wedge \sim Q$.
(e) $\sim(P \wedge Q)$ is equivalent to $P \Rightarrow \sim Q$.
(f) $P \wedge (Q \vee R)$ is equivalent to $(P \wedge Q) \vee (P \wedge R)$.
(g) $P \vee (Q \wedge R)$ is equivalent to $(P \vee Q) \wedge (P \vee R)$.

You will be asked to give a proof of this theorem (exercise 6). Before giving the proof you should think about the meaning behind each equivalence. For example, in (d), if $\sim(P \Rightarrow Q)$ is true, then this means $P \Rightarrow Q$ is false, which forces P to be true and Q to be false. But this means both P and $\sim Q$ are true, and so $P \wedge \sim Q$ is true. This reasoning can be reversed to show that if $P \wedge \sim Q$ is true, then $\sim(P \Rightarrow Q)$ is true. We conclude $\sim(P \Rightarrow Q)$ is true precisely when $P \wedge \sim Q$ is true, and thus they are equivalent. For example, "It is not the case that if triangle ABC has a right angle, then it is equilateral" is equivalent to "triangle ABC has a right angle and is not equilateral."

The final topic we consider in this section is the translation of sentences into propositional symbols. This is sometimes very complicated because English is such a rich and powerful language, with many nuances, and because the ambiguities we tolerate in English would destroy structure and usefulness if we allowed them in mathematics. Consider the sentence "The Dolphins won't make the playoffs unless the Rams win Sunday or the Bears win all the rest of their games." In conversation, the response would likely be "Explain that!," and in the course of the explanation we hope that the meaning would be made clear. As it stands, there are at least three nonequivalent ways in which people translate the sentence. The word "unless" is sometimes used to mean a conditional or its converse or a biconditional.

Examples. "That a set S is compact is sufficient for S to be bounded" is translated

$$S \text{ is compact} \Rightarrow S \text{ is bounded.}$$

"A necessary condition for a group G to be cyclic is that G is abelian" is translated

$$G \text{ is cyclic} \Rightarrow G \text{ is abelian.}$$

"A set S is infinite if S has an uncountable subset" is translated

$$S \text{ has an uncountable subset} \Rightarrow S \text{ is infinite.}$$

It should be noted that ordinary English is particularly flexible in using conditional connectives. For example, the sentence "I will pay you only if you fix my car" has the meaning of a biconditional.

If we let P denote the proposition "Roses are red," and Q denote the proposition "Violets are blue," we can translate the sentence "It is not the case that roses are red, nor that violets are blue" in at least two ways: $\sim(P \vee Q)$ or $\sim P \wedge \sim Q$. Fortunately these are equivalent by Theorem 1.2, part (c). Note that the proposition "Violets are purple" requires a new symbol, say R, since it expresses a new idea which cannot be formed from the components P and Q.

The sentence "17 and 35 have no common divisors" shows that the meaning, and not just the form of the sentence, must be considered in translating; it cannot be broken up into the two propositions: "17 has no common divisors" and "35 has no common divisors." Compare this with the proposition "17 and 35 have digits totaling 8," which can be written as a conjunction.

Example. "If x is an integer, then x is either even or odd" may be translated into $P \Rightarrow (Q \vee R)$, where P is "x is an integer," Q is "x is even," and R is "x is odd."

Example. "If p is a prime number that divides ab, then p divides a or b" becomes $(P \wedge Q) \Rightarrow (R \vee S)$, when P is "p is prime," Q is "p divides ab," R is "p divides a," and S is "p divides b."

Here are some phrases in English that are translated by using the connectives $\Rightarrow$ and $\Leftrightarrow$.

Use $P \Rightarrow Q$ to translate:

- If P, then Q.
- P implies Q.

- P is sufficient for Q.
- P only if Q.
- Q, if P.
- Q whenever P.
- Q is necessary for P.

Use $P \Leftrightarrow Q$ to translate:

- P if and only if Q.
- P is equivalent to Q.
- P is necessary and sufficient for Q.

The point of translating sentences into symbols is to help you recognize the structure of the sentence as an aid in determining its veracity or falsity. Since the statement is communicated through language, it is important that ambiguity be avoided. Frequently in mathematics this is accomplished by using a shorthand employing the symbols $\Rightarrow$, $\Leftrightarrow$, and so on.

The convention governing use of parentheses, adopted at the end of section 1.1, can be extended to the connectives $\Rightarrow$ and $\Leftrightarrow$. The connectives $\sim$, $\wedge$, $\vee$, $\Rightarrow$, and $\Leftrightarrow$ are applied in the order listed. That is, $\sim$ applies to the smallest possible proposition, and so forth. For example, $P \Rightarrow \sim Q \vee R \Leftrightarrow S$ is an abbreviation for $(P \Rightarrow ((\sim Q) \vee R)) \Leftrightarrow S$.

Exercises 1.2

1. Identify the antecedent and the consequent for each of the following conditional sentences.
   - ★ (a) If squares have three sides, then triangles have four sides.
   - (b) If the moon is made of cheese, then 8 is an irrational number.
   - (c) x divides 3 only if x divides 9.
   - ★ (d) f is differentiable is sufficient for f to be continuous.
   - (e) A sequence a is bounded whenever a is convergent.
   - ★ (f) A function f is bounded if f is integrable.
   - (g) $1 + 2 = 3$ is necessary for $1 + 1 = 2$.
2. ☆ Write the converse and contrapositive of each conditional sentence in exercise 1.
3. Which of the following conditional sentences are true?
   - ★ (a) If triangles have three sides, then squares have four sides.
   - (b) If a hexagon has six sides, then the moon is made of cheese.
   - ★ (c) If $7 + 6 = 14$, then $5 + 5 = 10$.
   - (d) If $5 < 2$ then $10 < 7$.
   - ★ (e) If one interior angle of a right triangle is $92°$, then the other interior angle is $88°$.
   - (f) If Euclid's birthday was April 2nd, then rectangles have four sides.
4. Which of the following are true?
   - ★ (a) Triangles have three sides iff squares have four sides.
   - (b) $7 + 5 = 12$ iff $1 + 1 = 2$.
   - ★ (c) x is even iff $x + 1$ is odd.
   - (d) $5 + 6 = 6 + 5$ iff $7 + 1 = 10$.
   - (e) A parallelogram has three sides iff 27 is prime.
5. Make truth tables for these propositions.
   - (a) $P \Rightarrow (Q \wedge P)$.
   - ★ (b) $(\sim P \Rightarrow Q) \vee (Q \Leftrightarrow P)$.

★ (c) $(\sim Q) \Rightarrow (Q \Leftrightarrow P)$.
(d) $((Q \Rightarrow S) \wedge (Q \Rightarrow R)) \Rightarrow ((P \vee Q) \Rightarrow (S \vee R))$.

☆ 6. Prove Theorem 1.2 by constructing truth tables for each equivalence.

7. Rewrite each of the following sentences using logical connectives.
★ (a) If f has a relative minimum at x_0 and if f is differentiable at x_0, then $f'(x_0) = 0$.
(b) If n is prime, then $n = 2$ or n is odd.
(c) A number p is real and not rational whenever p is irrational.
★ (d) If $x = 1$ or $x = -1$, then $|x| = 1$.
★ (e) f has a critical point at x_0 iff $f'(x_0) = 0$ or $f'(x_0)$ does not exist.
(f) S is compact iff S is closed and bounded.
(g) That bounded monotone sequences converge in S is necessary and sufficient for S to be complete.

8. Show the following pairs of statements are equivalent:
(a) $P \vee Q \Rightarrow R$ and $\sim R \Rightarrow \sim P \wedge Q$.
★ (b) $P \wedge Q \Rightarrow R$ and $P \wedge \sim R \Rightarrow \sim Q$.
(c) $P \Rightarrow Q \wedge R$ and $\sim Q \vee \sim R \Rightarrow \sim P$.
(d) $P \Rightarrow Q \vee R$ and $P \wedge \sim R \Rightarrow Q$.
(e) $(P \Rightarrow Q) \Rightarrow R$ and $(P \wedge \sim Q) \vee R$.
(f) $P \Leftrightarrow Q$ and $(\sim P \vee Q) \wedge (\sim Q \vee P)$.

9. Give, if possible, an example of a true implication $P \Rightarrow Q$ for which
★ (a) the converse is true.
(b) the converse is false.
(c) the contrapositive is true.
★ (d) the contrapositive is false.

10. Give, if possible, an example of a false implication $P \Rightarrow Q$ for which
(a) the converse is true.
(b) the converse is false.
(c) the contrapositive is true.
(d) the contrapositive is false.

11. For the sentences of exercise 7 (a), (b), (c), and (d), give the converse and contrapositive of each. Tell whether each converse and contrapositive is true or false.

SECTION 1.3. QUANTIFIERS

Unlike the propositions we have dealt with so far, the sentence $x \geq 3$ is neither true nor false. When the variable x is replaced by certain values (for example, 7) the resulting proposition is true while for other values of x (for example, 2) it is false. This is an example of an **open sentence**—that is, a sentence containing one or more variables—which becomes a proposition only when the variables are replaced by the names of particular objects. For notation, if an open sentence is called P and the variables are $x_1, x_2, \ldots, x_k$, we write $P(x_1, x_2, \ldots, x_k)$, and in the case of a single variable x, we write $P(x)$.

The sentence "x_1 is equal to $x_2 + x_3$" is an open sentence with three variables. If we denote this sentence by $P(x_1, x_2, x_3)$, then $P(7, 3, 4)$ is true, since $7 = 3 + 4$ while $P(1, 2, 3)$ is false.

The collection of objects that may be substituted to make an open sentence a true proposition is called the **truth set** of the sentence.† Before the truth set can be determined we must know what objects are available for consideration. That is, we must have specified a **universe** of discourse. In many cases the universe will be understood from the context. However, there

†The reader unfamiliar with the basic concepts of set theory is encouraged to read the first few pages of section 1 of chapter 2 before proceeding.

are times when it must be specified. For the sentence "$x < 5$," with the universe all the natural numbers, the truth set is $\{1, 2, 3, 4\}$ while, with the universe all integers, the truth set is $\{\ldots, -2, -1, 0, 1, 2, 3, 4\}$. For a sentence like "Some people dislike taxes," the universe is presumably the entire universe, which includes people, taxes, cows, triangles, and so forth.

Let $Q(x)$ be the sentence "$x^2 = 4$." With the universe specified as all real numbers, the truth set for $Q(x)$ is $\{2, -2\}$. With the universe the set of natural numbers, the truth set is $\{2\}$.

An open sentence $P(x)$ is not a proposition, but $P(a)$ is a proposition for any a in the universe. Another way to construct a proposition from $P(x)$ is to modify it with a quantifier.

Definitions. For an open sentence $P(x)$ with variable x, the sentence $(\forall x)P(x)$ is read "for all x, $P(x)$" and is true precisely when the truth set for $P(x)$ is the entire universe. The symbol $\forall$ is called the **universal quantifier.**

The sentence $(\exists x)P(x)$ is read "there exists x such that $P(x)$" and is true precisely when the truth set for $P(x)$ is nonempty. The symbol $\exists$ is called the **existential quantifier.**

If the universe is the set of all real numbers, then

- $(\exists x)(x \geq 3)$ is true and $(\forall x)(x \geq 3)$ is false;
- $(\exists x)(|x| > 0)$ is true and $(\forall x)(|x| > 0)$ is false;
- $(\exists x)(x^2 = -1)$ is false and $(\forall x)(x + 2 > x)$ is true.

We all agree that "All apples have spots" is quantified with $\forall$, but what form does it have? If we use $A(x)$ to represent "x is an apple" and $S(x)$ to represent "x has spots," and consider the universe to be the set of all fruits, should we write $(\forall x)(A(x) \wedge S(x))$ or $(\forall x)(A(x) \Rightarrow S(x))$? Let's examine both. First $(\forall x)(A(x) \wedge S(x))$ says "For all objects x in the universe, x both is an apple and has spots." Since our universe is not spotted apples, this is not the meaning we want. How about $(\forall x)(A(x) \Rightarrow S(x))$? This says "For all objects x, *if* x is an apple, then x has spots," which is the meaning we seek. In general a sentence of the form "All $P(x)$ are $Q(x)$" should by symbolized $(\forall x)(P(x) \Rightarrow Q(x))$.

Now let's consider "Some apples have spots." Will this be $(\exists x)(A(x) \wedge S(x))$ or $(\exists x)(A(x) \Rightarrow S(x))$? The first translation reads "There is an object x such that it is an apple and has spots" and thus $(\exists x)(A(x) \wedge S(x))$ is correct. On the other hand, $(\exists x)(A(x) \Rightarrow S(x))$ reads "There is an object x such that, if it is an apple, then it has spots." Thus $(\exists x)(A(x) \Rightarrow S(x))$ is not a correct symbolic translation, as it does not insure the existence of apples with spots; it only insures the existence of spots if there are any apples. In general a sentence of the form "Some $P(x)$ are $Q(x)$" should by symbolized $(\exists x)(P(x) \wedge Q(x))$.

Some statements have the meaning of a quantified sentence even when the words "for all" or "there exists" are not present. The sentence "If x is prime, then $x > 1$" is symbolically written $(\forall x)(x \text{ is prime} \Rightarrow x > 1)$. The sentence "Some real numbers have real multiplicative inverses" becomes $(\exists x)(x$ is a real number $\wedge (\exists y)(y$ is a real number $\wedge\, xy = 1))$. The sentence "Some people dislike taxes" becomes $(\exists x)(x$ is a person $\wedge (\forall y)(y$ is a tax $\Rightarrow x$ dislikes $y))$. The role of quantifiers, although vital, is not always emphasized when speaking mathematically. You should be on the alert for hidden quantifiers.

Let us consider the translation of "Some integers are even and some are odd." One correct translation is $(\exists x)(x$ is even$) \wedge (\exists x)(x$ is odd$)$, because the first quantifier $(\exists x)$ extends only as far as the word "even." After that, any variable (even x again) may be used to express "some are odd." It is equally correct and sometimes preferable to write $(\exists x)(x$ is even$) \wedge (\exists y)(y$ is odd$)$, but it would be wrong to write $(\exists x)(x$ is even $\wedge\, x$ is odd$)$.

You will be reminded in chapter 2 that the symbolism "$x \in A$" stands for "The object x is an element of the set A." Since this occurs so frequently in combinations with quantifiers, we adopt two abbreviations expressing the sentences "Every $x \in A$ has the property P" and "Some $x \in A$ has the property P." The first sentence could be restated as "If $x \in A$, then x has property P" and symbolized by $(\forall x)(x \in A \Rightarrow P(x))$. This is abbreviated to $(\forall x \in A)P(x)$. The second sentence above can be restated as "There is an x that is in A and that has property P" and symbolized by $(\exists x)(x \in A \wedge P(x))$. This is abbreviated by $(\exists x \in A)P(x)$.

Example. When $\mathbf{N}$ is the set of natural numbers and $\mathbf{R}$ is the set of real numbers, the sentence "For every natural number there is a real number greater than the natural number" may be symbolized by $(\forall n \in \mathbf{N})(\exists r \in \mathbf{R})(r > n)$.

We say two quantified sentences are *equivalent* iff they have the same truth set.

Theorem 1.3. If $A(x)$ is an open sentence with variable x, then

(a) $\sim(\forall x)A(x)$ is equivalent to $(\exists x)(\sim A(x))$.
(b) $\sim(\exists x)A(x)$ is equivalent to $(\forall x)(\sim A(x))$.

Proof.

(a) The sentence $\sim(\forall x)A(x)$ is true
iff $(\forall x)A(x)$ is false,
iff the truth set of $A(x)$ is not the universe,
iff the truth set of $\sim A(x)$ is nonempty,
iff $(\exists x)\sim A(x)$ is true.
Thus $\sim(\forall x)A(x)$ is true if and only if $(\exists x)\sim A(x)$ is true, so the propositions are equivalent.

(b) The proof of this part is an exercise (exercise 3). There is a proof similar to part (a) and another proof that uses part (a). ■

Example. Find a denial of "All primes are odd." Use the natural numbers as the universe.

The sentence may be symbolized $(\forall x)(x \text{ is prime} \Rightarrow x \text{ is odd})$. The negation is $\sim(\forall x)(x \text{ is prime} \Rightarrow x \text{ is odd})$, which by Theorem 1.3 (a) is equivalent to $(\exists x)(\sim(x \text{ is prime} \Rightarrow x \text{ is odd}))$. By Theorem 1.2 (d) this is equivalent to $(\exists x)(x \text{ is prime} \wedge \sim(x \text{ is odd}))$. Thus a denial is "There exists a prime number and it is not odd" or "Some prime number is even."

Example. Find a denial of "Every positive real number has a multiplicative inverse." Let the universe be the set of all real numbers.

The sentence is symbolized by $(\forall x)(x > 0 \Rightarrow (\exists y)(xy = 1))$. The negation is $\sim(\forall x)(x > 0 \Rightarrow (\exists y)(xy = 1))$, which may be rewritten successively as

$$(\exists x)\sim(x > 0 \Rightarrow (\exists y)(xy = 1)),$$
$$(\exists x)(x > 0 \wedge \sim(\exists y)(xy = 1)),$$
$$(\exists x)(x > 0 \wedge (\forall y)(xy \neq 1)).$$

In English, the last of these statements would read "There is a positive real number for which there is no multiplicative inverse."

Example. Let f be a function whose domain is the real numbers. Find a denial of the definition of "f is continuous at a."

Use the reals as the universe. A translation of the definition is

$$(\forall \epsilon)(\epsilon > 0 \Rightarrow (\exists \delta)(\delta > 0 \wedge (\forall x)(|x - a| < \delta \Rightarrow |f(x) - f(a)| < \epsilon))).$$

Make several applications of Theorems 1.2 and 1.3. (You are expected to carry out the steps.) A denial is

$$(\exists \epsilon)(\epsilon > 0 \wedge (\forall \delta)(\delta > 0 \Rightarrow (\exists x)(|x - a| < \delta \wedge |f(x) - f(a)| \geq \epsilon))).$$

In words, this denial is "There is a positive ϵ such that, for all positive δ, there exists x such that $|x - a| < \delta$, but $|f(x) - f(a)| \geq \epsilon$."

Example. Find a denial of "Some people dislike taxes." The sentence is symbolized by $(\exists x)(x \text{ is a person} \wedge (\forall y)(y \text{ is a tax} \Rightarrow x \text{ dislikes } y))$. By Theorems 1.2 and 1.3, its negation is equivalent to

$$(\forall x)(x \text{ is a person} \Rightarrow (\exists y)(y \text{ is a tax} \wedge x \text{ likes } y))$$

which is translated into "Everyone has a tax they like."

The last quantifier we shall consider is symbolized by $\exists!$.

> **Definition.** For an open sentence $P(x)$, the proposition $\mathbf{(\exists! x)P(x)}$ is read "There exists a unique x such that $P(x)$." The sentence $(\exists! x)P(x)$ is true when the truth set for $P(x)$ contains exactly one element from the universe.

In the universe of natural numbers, $(\exists! x)(x$ is an even positive prime) is true, since the truth set of "x is an even positive prime" contains only the number 2.

The sentence $(\exists! x)(x^2 = 4)$ is true when the universe is the natural numbers, and false when the universe is the integers.

Exercises 1.3

1. Translate the following English sentences into symbolic sentences with quantifiers. The universe for each is given in parentheses.
   - ★ (a) Not all precious stones are beautiful. (All stones)
   - (b) All precious stones are not beautiful. (All stones)
   - ★ (c) There is a smallest positive integer. (Real numbers)
   - (d) There is a rational number between any two real numbers. (Real numbers)
   - (e) Not all drunkards are truculent. (All people)
   - ★ (f) No one loves everybody. (All people)
   - (g) For every positive real number x, there is a unique real number y such that $2^y = x$. (Real numbers)
   - ★ (h) At least somebody cares about me. (All people)
   - (i) All people are honest or no one is honest. (All people)
   - (j) Some people are honest and some people aren't honest. (All people)
2. ☆ For each of the propositions in exercise 1, write a useful denial, and give an idiomatic English version.
3. ☆ Give two proofs of Theorem 1.3 (b).
4. Which of the following are true for the universe of all real numbers?
   - ★ (a) $(\forall x)(\exists y)(x + y = 0)$.
   - (b) $(\exists x)(\forall y)(x + y = 0)$.
   - (c) $(\exists x)(\exists y)(x^2 + y^2 = -1)$.
   - ★ (d) $(\forall x)(x > 0 \Rightarrow (\exists y)(y < 0 \land xy > 0))$.
   - (e) $(\forall y)(\exists x)(\forall z)(xy = xz)$.
   - ★ (f) $(\exists x)(\forall y)(x \leq y)$.
   - (g) $(\forall y)(\exists x)(x \leq y)$.
5. Give a denial of "You can fool some of the people all of the time and all of the people some of the time, but you cannot fool all of the people all of the time."
6. ★ Riddle: What is the English translation of the symbolic statement $\forall\exists\exists\forall$?

SECTION 1.4. MATHEMATICAL PROOFS

You have undoubtedly seen and written some proofs. The type of proofs you will be working with in this book are not like the proofs you remember from high school geometry, which may have conformed to a special format that included reasons for each step. A proof is a complete justification of the truth of a statement called a **theorem.** It generally begins with some hypotheses stated in the theorem and proceeds by correct reasoning to the claimed statement. Along the way it may draw upon other hypotheses, previously defined concepts, or some basic axioms setting forth properties of the concepts being considered. A **proof,** then, is a logically valid deduction of a theorem, from axioms or the theorem's premises, and may use previously proved theorems.

As you write a proof be sure it is not just a string of symbols. Every step of your proof should express a complete sentence. It is perfectly acceptable and at times advantageous to write the sentence symbolically, but you should be sure to include important connective words to complete the meaning of the symbols.

The validity of any statement can eventually be traced back to some initial set of concepts and assumptions. We cannot define all terms or else we would have a circular set of definitions. Neither can we prove all statements from previous ones. There must be an initial set of statements called **axioms** or **postulates** that are assumed true, and an initial set of concepts called **undefined terms,** from which new concepts can be introduced by means of **definitions** and new statements **(theorems)** can be deduced.

The validity of a proof is based on the notion of a **tautology**—that is, a proposition that is true for every assignment of truth values to its components. A few of the basic tautologies we shall refer to are

$$
\begin{array}{cr}
P \vee \sim P & \text{(Excluded Middle)} \\
P \Rightarrow Q \Leftrightarrow \sim Q \Rightarrow \sim P & \text{(Contrapositive)} \\
\left.\begin{array}{c} P \vee (Q \vee R) \Leftrightarrow (P \vee Q) \vee R \\ P \wedge (Q \wedge R) \Leftrightarrow (P \wedge Q) \wedge R \end{array}\right\} & \text{(Associativity)} \\
\left.\begin{array}{c} P \wedge (Q \vee R) \Leftrightarrow (P \wedge Q) \vee (P \wedge R) \\ P \vee (Q \wedge R) \Leftrightarrow (P \vee Q) \wedge (P \vee R) \end{array}\right\} & \text{(Distributivity)} \\
(P \Leftrightarrow Q) \Leftrightarrow [(P \Rightarrow Q) \wedge (Q \Rightarrow P)] & \\
\sim(P \Rightarrow Q) \Leftrightarrow P \wedge \sim Q & \\
\left.\begin{array}{c} \sim(P \wedge Q) \Leftrightarrow \sim P \vee \sim Q \\ \sim(P \vee Q) \Leftrightarrow \sim P \wedge \sim Q \end{array}\right\} & \text{(De Morgan's laws)} \\
P \Leftrightarrow (\sim P \Rightarrow Q \wedge \sim Q) & \text{(Contradiction)} \\
[(P \Rightarrow Q) \wedge (Q \Rightarrow R)] \Rightarrow (P \Rightarrow R) & \text{(Transitivity)} \\
P \wedge (P \Rightarrow Q) \Rightarrow Q & \text{(Modus Ponens)}
\end{array}
$$

Each may be verified as a tautology by its truth table, but you should also examine each to see that it expresses a logical relationship that is always true. See the discussion following Theorem 1.2.

In writing proofs, a working knowledge of tautologies is helpful because **a sentence whose symbolic translation is a tautology may be used at any time in a proof.** For example, if a proof involves a number x, one could at any time correctly assert "Either $x = 0$ or $x \neq 0$," since this is an instance of the tautology $P \vee \sim P$.

Most steps in a proof follow from earlier lines or other known results. **A statement Q may be deduced (asserted) in a proof if $(P_1 \wedge P_2 \wedge \ldots \wedge P_n) \Rightarrow Q$ is a tautology and $P_1, P_2, \ldots, P_n$ are either earlier statements in the proof, statements of previously proved theorems, hypotheses, or axioms.** We give several examples.

Example. If a proof contains the statement "x is a rational number and $x > 2$," we may deduce "x is a rational number" since $P \wedge Q \Rightarrow P$ is a tautology. Likewise, we may deduce "$x > 2$" since $P \wedge Q \Rightarrow Q$ is also a tautology.

Example. Suppose a proof contains the statements

$P_1 : f(5) = 0$ and $f(-2) = 0$.
$P_2 : f(7) = f(9)$.
$P_3 : f$ is a polynomial.

Then, because we are familiar with the Factor Theorem, we may use the statement

P_4: For a polynomial g, $g(a) = 0$ iff $x - a$ is a factor of g.

Then we may write as a new line in our proof the statement

$Q: x - 5$ is a factor of f.

This is because $(P_1 \wedge P_3 \wedge P_4) \Rightarrow Q$ is a tautology.

Example. If a proof contains the statement "x is an integer," we may deduce "x is a rational number," since it is known that "if x is an integer, then x is a rational number" and $P \wedge (P \Rightarrow Q) \Rightarrow Q$ is a tautology.

Example. The tautology $P \wedge (P \Leftrightarrow Q) \Rightarrow Q$ allows the **replacement** of any sentence P with an equivalent sentence Q. For instance, a proof containing the line "The product of real numbers a and b is zero" could later have the statement "a is zero or b is zero."

As another use of replacement, we may deduce the contrapositive of a conditional sentence in a proof. For instance, since we know "If $f'(x_0) = 0$, then f has a critical point at x_0," we could assert in a proof that "If f does not have a critical point at x_0, then $f'(x_0) \neq 0$."

Replacement can also be applied to components of statements in a proof. For instance, the statement "If x is an integer greater than 2 and x is a prime, then x does not divide 32" may be replaced by "If x is an odd prime, then x does not divide 32."

Example. We may *not* deduce the statement "n is a prime number" from the statements "If n is a prime number, then n is an integer" and "n is an integer," since this could only employ the proposition $(P \Rightarrow Q) \wedge Q \Rightarrow P$, which is not a tautology.

In a proof, you may at any time state any assumption or axiom. The introduction of an assumption generally takes the form "Assume P."

The strategy to construct a proof of a given theorem depends greatly on the logical form of the theorem's statement and on the particular concepts involved. As a general rule, when you write a step in a proof, ask yourself if making that assertion is valid in the sense that some tautology permits you to deduce it. It is not necessary to cite the tautology in your proof. In fact, with practice you should eventually come to write proofs without purposefully thinking about tautologies. What is necessary is that **every step must be justifiable.**

There is an allowance for great latitude for differences in taste and style among proof writers. Generally, the further you go in mathematics the less justification you will find given, because with more advanced topics more is expected of the reader. In this text our proofs will be complete and concise but **we shall on occasion insert parenthetical comments (offset by ⟨ ⟩ and in smaller type) to explain how and why a proof is proceeding as it is.** Comments set off by ⟨ ⟩ should not be taken as part of the proof but are inserted to help clarify the workings of the proof.

The first method of proof we will examine is the **direct proof** of a conditional sentence. How do you prove a sentence of the form $P \Rightarrow Q$? This implication is false only when P is true and Q is false, so it suffices to show this situation is an impossibility. The simplest way to proceed, then, is to **assume that P is true, and show (deduce) that Q is also true.** A direct proof of $P \Rightarrow Q$ will have the form

Proof.

Assume P.

⋮

Therefore Q.

Thus $P \Rightarrow Q$. ■

Example. Prove that if x is an odd integer, then $x + 1$ is even.

Proof. ⟨*The proof consists of assuming the antecedent and then writing down four equivalent statements, the last being the consequent.*⟩ Let x be an odd integer. Then $x = 2r + 1$ for some integer r. Thus $x + 1 = (2r + 1) + 1 = 2(r + 1)$ for some integer r. Since $r + 1$ is an integer, $x + 1$ is even. ■

Your strategy for a direct proof should include these steps:

1. Determine precisely the antecedent and consequent.
2. Replace (if necessary) the antecedent with a more usable equivalent.
3. Determine precisely what is to be shown.
4. Replace (if necessary) what is to be shown by something equivalent and more readily shown.
5. Develop a chain of statements, each deducible from its predecessors or other known results, that leads to what is to be shown.

Our second example requires recalling some facts about derivatives and limits of functions.

Example. Prove that, if $f'(x) = 5x$ and $g'(x) = 3$, then $(f + g)'(x) = 5x + 3$.

Proof. Assume that $f'(x) = 5x$ and $g'(x) = 3$. ⟨*Assume the antecedent.*⟩
Thus, $f'(x) = 5x$. ⟨$P \wedge Q \Rightarrow P$ *is a tautology.*⟩
So

$$\lim_{h \to 0} \frac{f(x + h) - f(x)}{h} = 5x \quad \langle \textit{Equivalent statement.} \rangle$$

Likewise, $g'(x) = 3$, so

$$\lim_{h \to 0} \frac{g(x + h) - g(x)}{h} = 3$$

⟨*The rest of the proof is mainly algebra and limit properties.*⟩ Therefore

$$\begin{aligned}
(f + g)'(x) &= \lim_{h \to 0} \frac{(f + g)(x + h) - (f + g)(x)}{h} \\
&= \lim_{h \to 0} \frac{f(x + h) + g(x + h) - f(x) - g(x)}{h} \\
&= \lim_{h \to 0} \left(\frac{f(x + h) - f(x)}{h} + \frac{g(x + h) - g(x)}{h} \right) \\
&= \lim_{h \to 0} \frac{f(x + h) - f(x)}{h} + \lim_{h \to 0} \frac{g(x + h) - g(x)}{h} = 5x + 3.
\end{aligned}$$

We conclude that $(f + g)'(x) = 5x + 3$. ■

Our final example of a direct proof is drawn from number theory. It makes use of the definition that the integer x **divides** the integer y iff $y = xn$ for some integer n.

Example. Prove that if a, b, c are integers, and a divides b, and a divides c, then a divides $b + c$.

Proof. Suppose a divides b and a divides c. ⟨*Now we use a more useful equivalent of "divides."*⟩ Then $b = an$ for some integer n and $c = am$ for some integer m. Thus $b + c = an + am = a(n + m)$. Since $n + m$ is an integer, ⟨*we use the fact that the sum of two integers is an integer,*⟩ a divides $b + c$. ■

Direct proofs of statements of the form $P \Rightarrow Q$ are not quite so straightforward when either P or Q is itself a compound proposition.

Example. To prove $P \Rightarrow (Q \vee R)$, one often proves either the equivalent $(P \wedge \sim Q) \Rightarrow R$ or the equivalent $(P \wedge \sim R) \Rightarrow Q$. For instance, in order to prove "If the polynomial f has degree 4, then f has a real zero or f can be written as the product of two irreducible quadratics," we would prove "If f has degree 4 and no real zeros, then f can be written as the product of two irreducible quadratics."

Example. To prove $(P \vee Q) \Rightarrow R$, one could proceed by cases, first proving $P \Rightarrow R$, then proving $Q \Rightarrow R$. This is valid because of the tautology $((P \vee Q) \Rightarrow R) \Leftrightarrow [(P \Rightarrow R) \wedge (Q \Rightarrow R)]$. The statement "If a quadrilateral has opposite sides equal or opposite angles equal, then it is a parallelogram" is proved by showing both "A quadrilateral with opposite sides equal is a parallelogram" and "A quadrilateral with opposite angles equal is a parallelogram."

Example. The usual method for proving a statement symbolized by $P \Rightarrow (Q \wedge R)$ would also be by cases. First show $P \Rightarrow Q$ and then show $P \Rightarrow R$. For instance, we would use this method to prove the statement "If two parallel lines are cut by a transversal, then corresponding angles are equal and alternate interior angles are equal."

A second form of proof for a conditional sentence is **proof by contrapositive.** The idea here is that since $P \Rightarrow Q$ is equivalent to its contrapositive, $\sim Q \Rightarrow \sim P$, we first give a direct proof of $\sim Q \Rightarrow \sim P$ and then conclude $P \Rightarrow Q$. This method works well when the connection between the denials of P and Q is easier to understand than the connection between P and Q themselves. The format of a proof by contrapositive is

> ***Proof.***
>
> Suppose $\sim Q$.
>
> ⋮
>
> Therefore, $\sim P$. (Via a direct proof)
> Thus $\sim Q \Rightarrow \sim P$.
> Therefore, $P \Rightarrow Q$. ■

Example. Prove that if m^2 is odd, then m is odd.

> ***Proof.*** Suppose m is not odd. ⟨*Suppose* $\sim Q$.⟩ Then m is even. Thus $m = 2k$ for some integer k. ⟨*Equivalent statement.*⟩ Then $m^2 = (2k)^2 = 4k^2 = 2(2k^2)$. Since m^2 is twice the integer $2k^2$, m^2 is even. ⟨*Deduce* $\sim P$.⟩ Thus, if m is even, then m^2 is even; so, by the contrapositive, if m^2 is odd, then m is odd. ■

A **proof by contradiction** makes use of the tautology $P \Leftrightarrow (\sim P \Rightarrow (Q \wedge \sim Q))$. In order to prove a proposition P, it is equivalent to prove $\sim P \Rightarrow (Q \wedge \sim Q)$ by a direct proof. Two aspects about this form of proof are especially noteworthy. First, this method of proof can be applied to any proposition P, whereas direct proofs and proofs by contrapositive can be used only for conditional sentences. Second, the proposition Q does not even appear on the left side of the tautology. The idea of proving $\sim P \Rightarrow (Q \wedge \sim Q)$ then has an advantage and a disadvantage. We don't know what proposition to use for Q, but any proposition that will do the job is a good one. This means a proof by contradiction will require a "spark of insight" to determine a useful Q. A proof by contradiction has the following form.

> ***Proof.***
>
> Suppose $\sim P$.
>
> ⋮
>
> Therefore, Q.
>
> ⋮
>
> Therefore, $\sim Q$.
> Hence $Q \wedge \sim Q$, a contradiction.
> Thus P. ■

Example. $\sqrt{2}$ is an irrational number.

> ***Proof.*** Suppose that $\sqrt{2}$ is a rational number. ⟨*Assume* $\sim P$.⟩ Then $\sqrt{2} = s/t$ where s and t are integers. Thus $2 = s^2/t^2$ and $2t^2 = s^2$. Since s^2 and t^2 are squares, s^2 contains an even number of 2's as factors, ⟨*This is our* Q *statement.*⟩ and t^2 contains an even number of 2's. But then $2t^2$ contains an odd number of 2's as factors. Since $s^2 = 2t^2$, s^2 has an odd number of 2's. ⟨*This is the statement* $\sim Q$.⟩ This is a contradiction. We conclude that $\sqrt{2}$ is irrational. ■

As a second example of a proof by contradiction we give a proof (attributed to Euclid) that there are an infinite number of primes.

Example. Prove that the set of primes is infinite.

Proof. Suppose the set of primes is finite. ⟨*Suppose* $\sim P$.⟩ Let the primes be $p_1, p_2, p_3, \ldots, p_k$ and consider the number $n = (p_1 p_2 \cdots p_k) + 1$. Since n is a natural number, n has a prime divisor q where $q > 1$. ⟨*The Q statement is* $q > 1$.⟩ Since q is a prime and $p_1, p_2, \ldots, p_k$ is the list of all primes, q is one of the p_i and thus q divides the product $p_1 p_2 \cdots p_k$. Since q also divides n, q divides $n - p_1 p_2 \cdots p_k$. ⟨*If a divides b and c, then a divides* $b - c$.⟩ But $n - p_1 p_2 \cdots p_k = 1$ and so $q = 1$. ⟨*This is* $\sim Q$.⟩ From this contradiction, we conclude that the set of primes is infinite. ■

Proofs of biconditional sentences are often based on the equivalence of $P \Leftrightarrow Q$ with $(P \Rightarrow Q) \wedge (Q \Rightarrow P)$. Many proofs of $P \Leftrightarrow Q$ will have the form:

> ***Proof.***
>
> (i) Show $P \Rightarrow Q$ by any method.
> (ii) Show $Q \Rightarrow P$ by any method.
>
> Therefore $P \Leftrightarrow Q$. ■

Of course, the two proofs in (i) and (ii) may use different methods. Frequently the proof of one part is more difficult than the other. This is true, for example, of the proof that "The natural number x is prime iff no positive integer greater than 1 and less than or equal to $\sqrt{x}$ divides x." It is obvious that "If x is prime, then no positive integer greater than 1 and less than or equal to $\sqrt{x}$ divides x," but the converse is a little more difficult to prove.

In some cases it is possible to prove a biconditional sentence $P \Leftrightarrow Q$ that uses the "iff" connective throughout. This amounts to starting with P and then replacing it with a sequence of equivalent statements, the last one being Q.

Example. Use the Law of Cosines to prove that a triangle with sides of length a, b, c is a right triangle with hypotenuse c if and only if $a^2 + b^2 = c^2$. See figure 1.1.

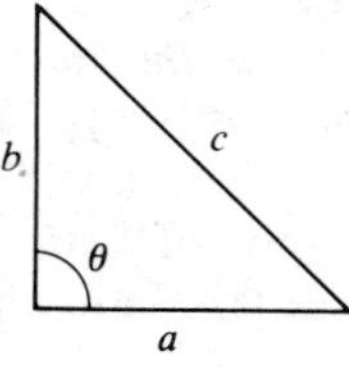

Figure 1.1

Proof. By the Law of Cosines, $a^2 + b^2 = c^2 - 2ab\cos\theta$. Thus

$$\begin{aligned} a^2 + b^2 = c^2 &\quad \text{iff} \\ 2ab\cos\theta = 0 &\quad \text{iff} \\ \cos\theta = 0 &\quad \text{iff} \\ \theta = 90^\circ &\quad \text{iff} \end{aligned}$$

the triangle is a right triangle with hypotenuse c. ■

One form of proof is known as **proof by exhaustion.** Such a proof consists of examining every possible case. This was our method in Theorem 1.1 where we examined all four cases, which were the four combinations of truth values for two propositions. Naturally, the idea of proof by exhaustion is appealing only when there is a small number of cases, or when large numbers of cases can be systematically handled, but there have been instances of very exhausting proofs involving great numbers of cases. In 1976, Kenneth Appel and Wolfgang Haken announced a proof of the long-standing Four-Color Conjecture. The original version of their proof contains 1879 cases, took 3½ years to develop, required 1200 hours of computer time, and is the result of over 10,000,000,000 calculations.

By now you may have the impression that given a set of axioms and definitions of a mathematical system, any correctly stated proposition in that system can be proved true or proved false. This is not the case. There are numerous examples in mathematics having **consistent** axiom systems (so that there exist structures satisfying all the axioms), which include statements, albeit true or false, that cannot be proved. Such statements are called **undecidable** in the system because their truth is independent of the truth of the axioms. The classic case of this involves the fifth of five postulates that Euclid (circa 300 B.C.) set forth as his basis for plane geometry: "Given a line and a point not on the line, exactly one line can be drawn through the point parallel to the line." For centuries some thought Euclid's axioms were not independent, believing that the fifth postulate could be proved from the other four. It was not until the nineteenth century that it became clear that the fifth postulate was undecidable. There are now theories of Euclidean geometry where the fifth postulate is assumed true and non-Euclidean geometries where it is assumed false. Both are perfectly reasonable areas for mathematical study.

Exercises 1.4

1. **Proofs to Grade.** Problems with this title throughout this book ask you to analyze an alleged proof of a claim and ask you to give one of three grades. Assign a grade of A (excellent) if the claim and proof are correct, even if the proof is not the simplest or the proof you would have given. Assign an E (failure) if the claim is incorrect, if the main idea of the proof is incorrect or if most of the statements in it are incorrect. Assign a grade of C (partial credit) for a proof that is largely correct, but contains one or two incorrect statements or justifications. Whenever the proof is incorrect, explain your grade. Tell what is incorrect and why.

★ (a) **Claim.** If m^2 is odd, then m is odd.
"Proof." Assume m is odd. Then $m = 2k + 1$ for some integer k. Therefore, $m^2 = (2k+1)^2 = 4k^2 + 4k + 1 = 2(2k^2 + 2k) + 1$, which is odd. Therefore, if m^2 is odd, then m is odd. ■

(b) **Claim.** If m^2 is odd, then m is odd.
"Proof." Assume that m^2 is not odd. Then m^2 is even and $m^2 = 2k$ for some integer k. Thus $2k$ is a perfect square; that is, $\sqrt{2k}$ is an integer. If $\sqrt{2k}$ is odd, then $\sqrt{2k} = 2n + 1$ for some integer n, which means
$m^2 = 2k = (2n+1)^2 = 4n^2 + 4n + 1 = 2(2k^2 + 2k) + 1$. Thus m^2 is odd, contrary to our assumption. Therefore $\sqrt{2k} = m$ must be even. Thus, if m^2 is not odd, then m is not odd. Hence, if m^2 is even, then m is even. ■

(c) **Claim.** If a, b, and c are integers and a divides both b and c, then a divides $b + c$.
"Proof." Assume that a does not divide $b + c$. Then there is no integer k such that $ak = b + c$. However, a divides b, so $am = b$ for some integer m, and a divides c, so $an = c$ for some integer n. Thus, $am + an = a(m + n) = b + c$. Therefore, $k = m + n$ is an integer satisfying $ak = b + c$. Thus, the assumption that a does not divide $b + c$ is false, and a does divide $b + c$. ■

★ (d) **Claim.** If t is an irrational number, then $5t$ is irrational.
"Proof." Suppose $5t$ is rational. Then $5t = p/q$ where p and q are integers and $q \neq 0$. Therefore $t = p/5q$ where p and $5q$ are integers and $5q \neq 0$, so t is rational. Therefore if t is irrational, then $5t$ is irrational. ■

SECTION 1.5. PROOFS INVOLVING QUANTIFIERS

Most theorems in mathematics are quantified sentences even though the quantifier may not actually appear in the statement. For example, proving "If x is an odd integer, then $x + 1$ is even," as we did in the last section, actually involves a quantifier, since the sentence has the symbolic translation $(\forall x)(x$ is odd $\Rightarrow x + 1$ is even). In this section, we present methods of proof for theorems of the forms $(\exists x)P(x)$, $(\forall x)P(x)$, and $(\exists! x)P(x)$. As in the previous section, the validity of the proof techniques is based upon employing statements that are always true.

There are several ways to prove **existence theorems**—that is, propositions of the form $(\exists x)P(x)$. The most direct, called a **constructive proof,** is to name or describe some object in the universe that actually makes $P(x)$ true. For example, stating "2 is a prime" is a proof of the theorem "There is a positive even prime integer." Here is another example.

Example. Prove the number $x = 4294967297$ is not a prime.

Proof. ⟨*It suffices to show the existence of natural numbers a and b both greater than 1, such that $x = ab$.*⟩ Léonard Euler first observed that
$x = (641)(6700417)$. Thus x is not prime. ■

It is also possible to give a proof of $(\exists x)P(x)$ by **contradiction.** The proof has the form:

> ***Proof.***
>
> Suppose $\sim(\exists x)P(x)$.
> Then $(\forall x)\sim P(x)$.
>
> ⋮
>
> Therefore, $Q \wedge \sim Q$, a contradiction.
> Hence, $\sim(\exists x)P(x)$ is false; so $(\exists x)P(x)$ is true. ■

The heart of such a proof is to prove a universally quantified statement.

Other proofs of existence theorems $(\exists x)P(x)$ show that there must be some object for which $P(x)$ is true, without ever actually producing a particular object. Both Rolles' Theorem and the Mean Value Theorem from your first course in calculus are good examples of this. Here is another.

Example. Prove the polynomial $r(x) = x^{71} - 2x^{39} + 5x - .3$ has a real zero.

> ***Proof.*** ⟨*What we must show has the form* $(\exists t)(r(t) = 0)$.⟩ By the Fundamental Theorem of Algebra, $r(x)$ has 71 zeros that are either real or complex. Since the polynomial has real coefficients, its complex zeros come in pairs. ⟨*By the Complex Root Theorem.*⟩ Hence, there are an even number of nonreal roots, and that leaves an odd number of real roots. Therefore $r(x)$ has at least one real root. ■

It is often necessary to prove that a statement of the form $(\forall x)P(x)$ is false. Since $\sim(\forall x)P(x)$ is equivalent to $(\exists x)\sim P(x)$, this amounts to a proof that $(\exists x)\sim P(x)$ is true. Any object t in the universe for which $\sim P(t)$ is true is called a **counterexample to $(\forall x)P(x)$.** For example, $f(x) = |x|$ is a counterexample to "Every function continuous at 0 is differentiable at 0." The number 2 is a counterexample to "All primes are odd." Some counterexamples have eluded mathematicians for centuries before being observed.

Finally, if an existentially quantified statement appears as a premise of a theorem, it should be assumed and then applied. For example, a proof of a statement of the form $(\exists x)P(x) \Rightarrow R$ could proceed as follows:

> ***Proof.***
>
> By hypothesis, $(\exists x)P(x)$.
> Let t be an object such that $P(t)$ is true.
>
> ⋮
>
> Therefore, R.
> Hence, $(\exists x)P(x) \Rightarrow R$. ■

In order to prove a proposition of the form $(\forall x)P(x)$, we must show that $P(x)$ is true for every object x in the universe. A direct proof of this is done by letting x represent an arbitrary object in the universe, and showing that $P(x)$ is true for that object without using any special properties of the object x. Then, since x is arbitrary, we can conclude that $(\forall x)P(x)$ is true.

Thus a **direct proof** of $(\forall x)P(x)$ has the form:

> ***Proof.***
>
> Let x be an arbitrary object in the universe. (Actually name the universe.)
>
> ⋮
>
> Hence, $P(x)$ is true.
> Since x is arbitrary, $(\forall x)P(x)$ is true. ■

The open sentence $P(x)$ will often be a combination of other open sentences and thus a deduction of $P(x)$ will require the selection of an appropriate proof technique.

Example. If x is an even integer, then x^2 is an even integer.

Proof. ⟨*The universe is evidently the integers. Here* $P(x)$ *denotes "x is even implies that* x^2 *is even."*⟩ Let x be an integer. Suppose x is even. ⟨*We are giving a direct proof of a conditional sentence.*⟩ Then $x = 2k$ for some integer k. Therefore $x^2 = (2k)^2 = 2(2k^2)$. Since $2k^2$ is an integer, x^2 is even. Hence, if x is even, then x^2 is even. ■

Another method of proof of a statement of the form of $(\forall x)P(x)$ is by **contradiction.** The form is as follows:

> ***Proof.***
>
> Suppose $\sim(\forall x)P(x)$.
> Then, $(\exists x)\sim P(x)$.
> Let t be an object such that $\sim P(t)$.
>
> ⋮
>
> Therefore, $Q \wedge \sim Q$.
> Thus, $(\exists x)\sim P(x)$ is false; so its denial $(\forall x)P(x)$ is true. ■

This is essentially a proof of an existential statement amenable to the techniques given earlier in this section.

The last proof method we give is for the $\exists!$ quantifier. There is a standard technique for proving a proposition of the form $(\exists! x)P(x)$.

Proof.

(i) Prove $(\exists x)P(x)$ is true by any method.
(ii) Assume that t_1 and t_2 are objects in the universe such that $P(t_1)$ and $P(t_2)$ are true.

⋮

Therefore, $t_1 = t_2$.

We conclude $(\exists! x)P(x)$. ■

Example. Prove that the polynomial $r(x) = x - 3$ has a unique zero.

Proof.

(i) Since $r(3) = 3 - 3 = 0$, 3 is a zero of $r(x)$.
(ii) Suppose that t_1 and t_2 are two zeros of $r(x)$. Then $r(t_1) = 0 = r(t_2)$. Therefore, $t_1 - 3 = t_2 - 3$. Thus, $t_1 = t_2$. Therefore, $r(x) = x - 3$ has a unique zero. ■

Example. Every nonzero real number has a unique multiplicative inverse.

Proof. ⟨*The theorem is symbolized* $(\forall x)(x \neq 0 \Rightarrow (\exists! y)($*y is real and* $xy = 1))$.⟩ Let x be a real number, ⟨*We start this way because the statement to be proved is quantified with* $\forall$ ⟩, and suppose $x \neq 0$. ⟨*Assume the antecedent.*⟩ We must show that $xy = 1$ for exactly one real number y.

(i) Let $y = 1/x$. ⟨*This is a constructive proof that there exists an inverse.*⟩ Then y is a real number since $x \neq 0$, and $xy = x(1/x) = x/x = 1$. Thus, x has a multiplicative inverse.
(ii) Now suppose y and z are two real multiplicative inverses for x. ⟨*This* y *is not necessarily the* $y = 1/x$ *in part* (i).⟩ Then $xy = 1$ and $xz = 1$. Thus, $xy = xz$ and $xy - xz = x(y - z) = 0$. Since $x \neq 0$, $y - z = 0$. Therefore $y = z$.

Thus, every nonzero real number has a unique multiplicative inverse. ■

Great care must be taken in proofs that contain expressions involving more than one quantifier. Here are some manipulations of quantifiers that permit valid deductions.

1. $(\forall x)(\forall y)P(x, y) \Leftrightarrow (\forall y)(\forall x)P(x, y)$.
2. $(\exists x)(\exists y)P(x, y) \Leftrightarrow (\exists y)(\exists x)P(x, y)$.
3. $((\forall x)P(x) \vee (\forall x)Q(x)) \Rightarrow (\forall x)(P(x) \vee Q(x))$.
4. $(\forall x)(P(x) \Rightarrow Q(x)) \Rightarrow ((\forall x)P(x) \Rightarrow (\forall x)Q(x))$.
5. $(\forall x)(P(x) \wedge Q(x)) \Leftrightarrow ((\forall x)P(x) \wedge (\forall x)Q(x))$.
6. $(\exists x)(\forall y)P(x, y) \Rightarrow (\forall y)(\exists x)P(x, y)$.

You should convince yourself that each of these is a logically valid conditional or biconditional. For example, the last on the list is always true because, if $(\exists x)(\forall y)P(x, y)$ is true, then there is (at least) one x that makes $P(x, y)$ true no matter what y is. Therefore, for any y, $(\exists x)P(x, y)$ is true because the one x exists.

It is equally important to be aware of the most common **incorrect deductions** making use of quantifiers. We list four here and give counterexamples.

1. $(\exists x)P(x) \Rightarrow (\forall x)P(x)$ is not valid.
   If the universe is all integers and $P(x)$ is the sentence "x is odd," then $P(5)$ is true and $P(8)$ is false. Thus, $(\exists x)P(x)$ is true and $(\forall x)P(x)$ is false, so the implication fails.
2. $(\forall x)(P(x) \vee Q(x)) \Rightarrow (\forall x)P(x) \vee (\forall x)Q(x)$ is not valid.
   We use the same example as in 1 and, in addition, let $Q(x)$ be "x is even." Then it is true that "All integers are either odd or even" but false that "Either all integers are odd or all integers are even."
3. $((\forall x)P(x) \Rightarrow (\forall x)Q(x)) \Rightarrow (\forall x)(P(x) \Rightarrow Q(x))$ is not valid.
   The same example in 2 can be used here. Because $(\forall x)P(x)$ is false, $(\forall x)P(x) \Rightarrow (\forall x)Q(x)$ is true. However, $(\forall x)(P(x) \Rightarrow Q(x))$ is false.
4. $(\forall y)(\exists x)P(x, y) \Rightarrow (\exists x)(\forall y)P(x, y)$ is not valid.
   This is probably the most troublesome of all the possibilities for dealing with quantifiers. Let the universe be the set of all married people and $P(x, y)$ be the sentence "x is married to y." Then, $(\forall y)(\exists x)P(x, y)$ is true, since everyone is married to someone. But $(\exists x)(\forall y)P(x, y)$ would be translated as "There is some married person who is married to every married person," which is clearly false.

Exercises 1.5

1. **Proofs to Grade.**

★ (a) **Claim.** Every polynomial of degree 3 with real coefficients has a real zero.
"Proof." We note that the polynomial $p(x) = x^3 - 8$ has degree 3, real coefficients, and a real zero ($x = 2$). Thus, the statement "Every polynomial of degree 3 with real coefficients does not have a real zero" is false, and hence its denial, "Every polynomial of degree 3 with real coefficients has a real zero," is true. ■

★ (b) **Claim.** There is a unique polynomial whose first derivative is $2x + 3$ and which has a zero at $x = 1$.
"Proof." The antiderivative of $2x + 3$ is $x^2 + 3x + C$. If we let $p(x) = x^2 + 3x - 4$, then $p'(x) = 2x + 3$ and $p(1) = 0$. So $p(x)$ is the desired polynomial. ■

(c) **Claim.** There exists an integer x such that $x + 13$ is a perfect square.
"Proof." Let the universe be the set of all integers x such that $x + 13$ is a perfect square. Then the following sentence, $(\forall x)(x + 13$ is a perfect square), is true. Thus, we may deduce the sentence $(\exists x)(x + 13$ is a perfect square) as true. ■

★ (d) **Claim.** There exists an irrational number r such that $r^{\sqrt{2}}$ is rational.
"Proof." If $\sqrt{3}^{\sqrt{2}}$ is rational, then $r = \sqrt{3}$ is the desired example. Otherwise, $\sqrt{3}^{\sqrt{2}}$ is irrational and $(\sqrt{3}^{\sqrt{2}})^{\sqrt{2}} = (\sqrt{3})^2 = 3$, which is rational. Therefore, either $\sqrt{3}$ or $\sqrt{3}^{\sqrt{2}}$ is an irrational number r such that $r^{\sqrt{2}}$ is rational. ■

(e) **Claim.** Every real function is continuous at $x = 0$.
"Proof." Since either a sentence or its negation is true, we know that for every real function either it is continuous at $x = 0$ or it is not continuous at $x = 0$. Thus, for every real function it is continuous at $x = 0$ or for every real function it is not

continuous at $x = 0$. The latter half of this sentence is false, since $f(x) = x^2$ *is* a real function that is continuous at $x = 0$. Since the sentence is a disjunction, the first half is true. Thus, for every real function, it is continuous at $x = 0$. ■

(f) **Claim.** If x is a prime, then $x + 7$ is composite.
"Proof." Let x be a prime number. If $x = 2$, then $x + 7 = 9$, which is composite. If $x \neq 2$, then x is odd, so $x + 7$ is even and greater than 2. In this case, too, $x + 7$ is composite. Therefore, if x is prime, then $x + 7$ is composite. ■

(g) **Claim.** For all irrational numbers t, $t - 8$ is irrational.
"Proof." Suppose there exists an irrational number t such that $t - 8$ is rational. Then $t - 8 = p/q$, where p and q are integers and $q \neq 0$. Then $t = p/q + 8 = (p + 8q)/q$, with $p + 8q$ and q integers and $q \neq 0$. This is a contradiction, because t is irrational. Therefore, for all irrational numbers t, $t - 8$ is irrational. ■

Set Theory 2

Much of mathematics is written in terms of sets. We assume that you have had some contact with the basic notions of sets, unions, and intersections. Sections 1 and 2 consist of a brief review of sets and operations and present the notation we shall use. Section 2 provides the opportunity for you to prove some set-theoretic results. In section 3, we will extend the set operations of union and intersection and encounter indexed collections of sets. Section 4 deals with inductive sets and proof by induction.

SECTION 2.1. BASIC NOTIONS OF SET THEORY

Instead of defining what a set is, we shall informally understand a **set** to be any specified collection of objects. The objects in a given set are called **elements** (or members) of the set. Most important, given an object and a set, one must be able to determine whether the object is an element of that set.

In general, capital letters are used to denote sets and lower case letters to denote objects. If the object x is an element of set A, we write $x \in A$; if not—that is, if $\sim(x \in A)$—we write $x \notin A$. For example, if B is the set of all baseball players who have hit at least 700 home runs, we write Henry Aaron $\in B$ and Joe Slobotnik $\notin B$.

Sets can be described in words, such as "the set of odd integers between 0 and 12," or the elements may be listed, as in $\{1, 3, 5, 7, 9, 11\}$, or even partially listed, as in $\{1, 3, 5, \ldots, 11\}$. Such explicit descriptions of sets are often impractical and sometimes impossible. To designate most sets we will use the following notation,

$$\{x: P(x)\},$$

where $P(x)$ is an open sentence description of the property that defines the set. The variable x in $\{x: P(x)\}$ is a dummy variable in the sense that any letter or symbol serves equally well. If $P(x)$ is the sentence "x is an odd integer between 0 and 12," then $\{x: P(x)\} = \{y: P(y)\} = \{1, 3, 5, \ldots, 11\}$.

A bit of caution should be observed in the type of property used in this notation. It is not true that for every open sentence $P(x)$, there corresponds a set $\{x: P(x)\}$. For more on this, see exercise 15.

Special notation will be used for the following sets of numbers:

$\mathbf{N} = \{1, 2, 3, \ldots\}$, the natural numbers
$\mathbf{Z} = \{\ldots, -3, -2, -1, 0, 1, 2, 3, \ldots\}$, the integers
$\mathbf{Q} =$ the set of rational numbers
$\mathbf{R} =$ the set of real numbers.

In addition, for $a, b \in \mathbf{R}$ with $a < b$, we will use

$$[a, b] = \{x : x \in \mathbf{R} \text{ and } a \leq x \leq b\}$$

and

$$(a, b) = \{x : x \in \mathbf{R} \text{ and } a < x < b\}$$

to represent the **closed** (and **open,** respectively) **interval from *a* to *b*.** The **half-open** (or **half-closed**) intervals $[a, b)$ and $(a, b]$ are defined similarly. Also,

$$(a, \infty) = \{x : x \in \mathbf{R} \text{ and } a < x\}$$

and

$$(-\infty, a) = \{x : x \in \mathbf{R} \text{ and } x < a\}$$

will be called **open rays,** while

$$[a, \infty) = \{x : x \in \mathbf{R} \text{ and } a \leq x\}$$

and

$$(-\infty, a] = \{x : x \in \mathbf{R} \text{ and } x \leq a\}$$

are called **closed rays.**

One should be careful not to confuse $(1, 6)$ with $\{2, 3, 4, 5\}$ since $(1, 6)$ is defined as **all real** numbers between 1 and 6 and contains, for example, 2, π, 3, $\log(15)$, and $\frac{11}{5}$.

Definition. Let $\varnothing = \{x : x \neq x\}$. Then $\varnothing$ is a set with no elements, and is called an **empty set.**

The sentence $x \in \varnothing$ is false for every object x. We shall soon see that there can be only one empty set.

Definition. Let A and B be sets. We say ***A* is a subset of *B*** iff every element of A is also an element of B. In symbols, this is

$$A \subseteq B \Leftrightarrow (\forall x)(x \in A \Rightarrow x \in B).$$

Since the statement $A \subseteq B$ is symbolized with the quantifier $(\forall x)$, a direct proof of $A \subseteq B$ begins with the words "Let x be any object." We continue the proof by showing that $x \in A \Rightarrow x \in B$ for that object x.

Theorem 2.1. For any set A, $\varnothing$ is a subset of A.

> ***Proof.*** Let A be any set. We must show that $(\forall x)(x \in \varnothing \Rightarrow x \in A)$. Let x be any object. Because any conditional sentence is true when the antecedent is false, $(x \in \varnothing \Rightarrow x \in A)$ is true. Therefore, $\varnothing \subseteq A$. ■

Theorem 2.2. For any set B, $B \subseteq B$.

> ***Proof.*** Let B be any set. To prove $B \subseteq B$, we must show that, for all objects x, if $x \in B$, then $x \in B$. Let x be any object. Then $x \in B \Rightarrow x \in B$ is true. ⟨*This is an instance of the tautology* $P \Rightarrow P$.⟩ Therefore, $(\forall x)(x \in B \Rightarrow x \in B)$, and so $B \subseteq B$. ■

Theorem 2.3. If $A \subseteq B$ and $B \subseteq C$, then $A \subseteq C$.

> ***Proof.*** Exercise 10. ■

If A is not a subset of B, we write $A \not\subseteq B$. A denial of $A \subseteq B$ is the proposition $(\exists x)(x \in A$ and $x \notin B)$. Thus, in order to prove $A \not\subseteq B$, all that is required is to show that some element of A is not an element of B. For example, $\{1, 5, 9, 6\} \not\subseteq \{2, 7, 6\}$ because $5 \in \{1, 5, 9, 6\}$ and $5 \notin \{2, 7, 6\}$.

For a given set A, the subsets $\varnothing$ and A are called **improper subsets** of A, while any subset of A other than $\varnothing$ or A is called a **proper subset.**

Definition. Let A be a set. The **power set of A** is the set whose elements are the subsets of A and is denoted $\mathscr{P}(A)$. Thus,

$$\mathscr{P}(A) = \{B : B \subseteq A\}.$$

Example. Let $A = \{a, b, c\}$. Then
$\mathscr{P}(A) = \{\varnothing, \{a\}, \{b\}, \{c\}, \{a, b\}, \{a, c\}, \{b, c\}, A\}$.

It should be noted that the elements of the set $\mathscr{P}(A)$ are themselves sets, specifically the subsets of A. Also, in working with sets whose elements are sets, it is important to recognize the distinction between "is an element of" and "is a subset of." To use $A \in B$ correctly, we must consider whether the object A (which happens to be a set) is an element of the set B, whereas $A \subseteq B$ requires determining whether all objects in the set A are also in B.

Let $A = \{a, b, c\}$. Then $a \in A$, $b \in A$, $\{a, b\} \subseteq A$, and $\{a, b\} \in \mathscr{P}(A)$. Notice that $\{a, b\}$ is an element of $\mathscr{P}(A)$ because $\{a, b\}$ is a subset of A. Also, $\{c\} \notin A$ but $\{c\} \in \mathscr{P}(A)$.

Let $X = \{\{1, 2, 3\}, \{4, 5\}, 6\}$. Then X is a set with three elements, namely, the set $\{1, 2, 3\}$, the set $\{4, 5\}$, and the number 6. The set $\{\{4, 5\}\}$ has one element; it is $\{4, 5\}$. All of the following are true: $\{4\} \subseteq \{4, 5\}$, $\{4, 5\} \in X$, $\{4, 5\} \not\subseteq X$, $6 \in X$, $\{6\} \subseteq X$, and $\mathscr{P}(X) = \{\varnothing, \{\{1, 2, 3\}\}, \{\{4, 5\}\}, \{6\}, \{\{1, 2, 3\}, \{4, 5\}\}, \{\{1, 2, 3\}, 6\}, \{\{4, 5\}, 6\}, X\}$.

Notice that for the set $A = \{a, b, c\}$ in the example above, A has three elements and $\mathscr{P}(A)$ has $2^3 = 8$ elements. As we see in the next theorem, this observation can be generalized to any set of n elements. It is for this reason that 2^A is often used to denote the power set of A.

Theorem 2.4. If A is a finite† set with n elements, then $\mathscr{P}(A)$ has 2^n elements.

Proof. ⟨*The number of elements in* $\mathscr{P}(A)$ *is the number of subsets of* A. *Thus to prove this result we must count all of the subsets of* A.⟩ Suppose $A = \{x_1, x_2, \ldots, x_n\}$. In order to describe a subset B of A, we need to know for each $x_i \in A$ whether the element is in B. For each x_i, there are two possibilities ($x_i \in B$ or $x_i \notin B$), so there are $2 \cdot 2 \cdot 2 \cdot \cdots \cdot 2$ (n factors) different ways of making a subset of A. Therefore, $\mathscr{P}(A)$ has 2^n elements. ■

Theorem 2.5. Let A and B be sets. Then $A \subseteq B$ iff $\mathscr{P}(A) \subseteq \mathscr{P}(B)$.

Proof. ⟨*This is a good example of a two-part proof of a biconditional, which is easier than an iff proof.*⟩

(i) We must show that $A \subseteq B$ implies $\mathscr{P}(A) \subseteq \mathscr{P}(B)$. Assume that $A \subseteq B$, and let $X \in \mathscr{P}(A)$. We must show that $X \in \mathscr{P}(B)$. But $X \in \mathscr{P}(A)$ implies $X \subseteq A$. Since $X \subseteq A$ and $A \subseteq B$, then $X \subseteq B$ by Theorem 2.3. But $X \subseteq B$ implies $X \in \mathscr{P}(B)$. Therefore, $X \in \mathscr{P}(A)$ implies $X \in \mathscr{P}(B)$. Thus, $\mathscr{P}(A) \subseteq \mathscr{P}(B)$.

(ii) We must show that $\mathscr{P}(A) \subseteq \mathscr{P}(B)$ implies $A \subseteq B$. Assume that $\mathscr{P}(A) \subseteq \mathscr{P}(B)$. By Theorem 2.2, $A \subseteq A$; so $A \in \mathscr{P}(A)$. Since $\mathscr{P}(A) \subseteq \mathscr{P}(B)$, $A \in \mathscr{P}(B)$. Therefore, $A \subseteq B$. ■

The second half of the proof above could have been done by showing directly that if $x \in A$, then $x \in B$. Such proofs are often called element-chasing proofs. The given proof is preferable, because it makes use of a theorem we already know. Both proofs are correct. When you write proofs you may choose one method of proof over another because it is shorter, or easier to understand, or for any other reason.

We have seen that a set may be described in different ways. Often it becomes necessary to know whether two descriptions of sets do in fact yield the same set. Intuitively, two sets A and B are "equal" if they contain exactly the same elements. We might say, then, that

$$A = B \text{ iff } (\forall x)(x \in A \Leftrightarrow x \in B).$$

†The formal definition of finite appears in chapter 5.

However, this is equivalent to

$$(\forall x)(x \in A \Rightarrow x \in B \text{ and } x \in B \Rightarrow x \in A),$$

which is longer but more natural to use in proving that $A = B$.

Definition. Let A and B be sets. Then $\boldsymbol{A = B}$ iff $A \subseteq B$ and $B \subseteq A$.

According to this definition, the task of proving an equality between two sets A and B is accomplished by showing that each set is a subset of the other. Occasionally, $A = B$ can be proved by showing that $x \in A$ iff $x \in B$.

Example. Prove that $X = Y$, where $X = \{x: x \text{ is a solution to } x^2 - 1 = 0\}$ and $Y = \{-1, 1\}$.

Proof. We must show (i) $Y \subseteq X$ and (ii) $X \subseteq Y$.

(i) We show $Y \subseteq X$ by individually checking each element of Y. By substitution, we see that both 1 and -1 are solutions to $x^2 - 1 = 0$. Thus, $Y \subseteq X$.

(ii) Next, we must show $X \subseteq Y$. Let $t \in X$. Then, by definition of X, t is a solution to $x^2 - 1 = 0$. Thus, $t^2 - 1 = 0$. Factoring, we have $(t - 1)(t + 1) = 0$. This product is 0 exactly when $t - 1 = 0$ or $t + 1 = 0$. Therefore $t = 1$ or $t = -1$. Thus, if t is a solution, then $t = 1$ or $t = -1$; so $t \in Y$. This proves $X \subseteq Y$.

By (i) and (ii), $X \subseteq Y$ and $Y \subseteq X$; so $X = Y$. ■

We are now in a position to prove that there is only one empty set.

Theorem 2.6. If A and B are sets with no elements, then $A = B$.

Proof. Since A is empty, the sentence $(\forall x)(x \in A \Rightarrow x \in B)$ is true. Therefore $A \subseteq B$. Similarly $(\forall x)(x \in B \Rightarrow x \in A)$ is true, so $B \subseteq A$. There, by definition of set equality, $A = B$. ■

Exercises 2.1

1. Write the following sets by using the set notation $\{x: P(x)\}$.
   ★ (a) The set of natural numbers strictly less than 6.
   (b) The set of integers whose square is less than 17.
   ★ (c) $[2, 6]$
   (d) $(-1, 9]$
   (e) $[-5, -1)$
   (f) The set of rational numbers less than -1.

☆ 2. Write each of the sets in exercise 1 by listing (if possible) all of its elements.

3. True or false?
★ (a) $\mathbf{N} \subseteq \mathbf{Q}$
(b) $\mathbf{Q} \subseteq \mathbf{Z}$
★ (c) $\mathbf{N} \subseteq \mathbf{R}$
(d) $[\frac{1}{2}, \frac{5}{2}] \subseteq \mathbf{Q}$
★ (e) $[\frac{1}{2}, \frac{5}{2}] \subseteq (\frac{1}{2}, \frac{5}{2})$
(f) $\mathbf{R} \subseteq \mathbf{Q}$
★ (g) $[7, 10] \subseteq \mathbf{R}$
(h) $[2, 5] = \{2, 3, 4, 5\}$
★ (i) $[7, 10) \subseteq \{7, 8, 9, 10\}$
(j) $(6, 9] \subseteq [6, 10)$
4. Write the power set, $\mathscr{P}(X)$, for each of the following sets:
★ (a) $X = \{0, \Delta, \Box\}$
(b) $X = \{S, \{S\}\}$
★ (c) $X = \{\varnothing, \{a\}, \{b\}, \{a, b\}\}$
(d) $X = \{1, \{2, \{3\}\}\}$
(e) $X = \{1, 2, 3, 4\}$
5. List all of the proper subsets for each of the following sets:
★ (a) $\varnothing$
(b) $\{1\}$
★ (c) $\{1, 2\}$
(d) $\{\{\varnothing\}\}$
(e) $\{\varnothing, \{\varnothing\}\}$
(f) $\{0, \Delta, \Box\}$
6. True or false?
★ (a) $\varnothing \in \{\varnothing, \{\varnothing\}\}$
(b) $\varnothing \subseteq \{\varnothing, \{\varnothing\}\}$
★ (c) $\{\varnothing\} \in \{\varnothing, \{\varnothing\}\}$
(d) $\{\varnothing\} \subseteq \{\varnothing, \{\varnothing\}\}$
★ (e) $\{\{\varnothing\}\} \in \{\varnothing, \{\varnothing\}\}$
(f) $\{\{\varnothing\}\} \subseteq \{\varnothing, \{\varnothing\}\}$
★ (g) For every set A, $\varnothing \in A$
(h) For every set A, $\{\varnothing\} \subseteq A$
★ (i) $\{\varnothing, \{\varnothing\}\} \subseteq \{\{\varnothing, \{\varnothing\}\}\}$
(j) $\{1, 2\} \in \{\{1, 2, 3\}, \{1, 3\}, 1, 2\}$
★ (k) $\{1, 2, 3\} \subseteq \{1, 2, 3, \{4\}\}$
(l) $\{\{4\}\} \subseteq \{1, 2, 3, \{4\}\}$
7. If possible, give an example of each of the following:
★ (a) Sets A, B, and C such that $A \subseteq B$, $B \not\subseteq C$, and $A \subseteq C$
(b) Sets A, B, and C such that $A \subseteq B$, $B \subseteq C$, and $C \subseteq A$
★ (c) Sets A, B, and C such that $A \not\subseteq B$, $B \not\subseteq C$, and $A \subseteq C$
(d) Sets A and B such that $A \subseteq B$, and $\mathscr{P}(B) \subseteq \mathscr{P}(A)$
★ (e) A set A such that $\mathscr{P}(A) = \varnothing$
(f) Sets A, B, and C such that $A \not\subseteq B$, $B \subseteq C$, and $\mathscr{P}(A) \subseteq \mathscr{P}(C)$
8. True or false?
★ (a) $\varnothing \in \mathscr{P}(\{\varnothing, \{\varnothing\}\})$
(b) $\{\varnothing\} \in \mathscr{P}(\{\varnothing, \{\varnothing\}\})$
★ (c) $\{\{\varnothing\}\} \in \mathscr{P}(\{\varnothing, \{\varnothing\}\})$
(d) $\varnothing \subseteq \mathscr{P}(\{\varnothing, \{\varnothing\}\})$
★ (e) $\{\varnothing\} \subseteq \mathscr{P}(\{\varnothing, \{\varnothing\}\})$
(f) $\{\{\varnothing\}\} \subseteq \mathscr{P}(\{\varnothing, \{\varnothing\}\})$
★ (g) $3 \in \mathbf{Q}$
(h) $\{3\} \subseteq \mathscr{P}(\mathbf{Q})$
★ (i) $\{3\} \in \mathscr{P}(\mathbf{Q})$
(j) $\{\{3\}\} \subseteq \mathscr{P}(\mathbf{Q})$
★ (k) $\{3\} \subseteq \mathbf{Q}$
(l) $\{\{3\}\} \in \mathscr{P}(\mathbf{Q})$
9. Let $A = \{x: P(x)\}$ and $B = \{x: Q(x)\}$.
★ (a) Prove that if $(\forall x)(P(x) \Rightarrow Q(x))$, then $A \subseteq B$.
(b) Prove that if $(\forall x)(P(x) \Leftrightarrow Q(x))$, then $A = B$.
10. Prove Theorem 2.3.
☆ 11. Prove that if $A \subseteq B$, $B \subseteq C$, and $C \subseteq A$, then $A = B$ and $B = C$.
12. Prove that $X = Y$, where $X = \{x: x \in \mathbf{R}$ and x is a solution to $x^2 - 7x + 12 = 0\}$ and $Y = \{3, 4\}$.
13. Prove that $X = Y$, where $X = \{x \in \mathbf{Z}: |x| \leq 3\}$ and $Y = \{-3, -2, -1, 0, 1, 2, 3\}$.
14. Prove that $X = Y$, where $X = \{x \in \mathbf{N}: x^2 < 30\}$ and $Y = \{1, 2, 3, 4, 5\}$.
★ 15. (Russell paradox.) A logical difficulty arises from the idea, which at first appears natural, of calling any collection of objects a set. A set B is **ordinary** if $B \notin B$. For example, if B is the set of all chairs, then $B \notin B$, for B is not a chair. It is only in the case of very unusual collections that we are tempted to say that a set is a member of itself. (The collection of all abstract ideas certainly is an abstract idea.) Let $X = \{x: x$ is an ordinary set$\}$. Is $X \in X$? Is $X \notin X$? What should we say about the collection of all ordinary sets?

16. **Proofs to Grade.** (See the instruction for problem 1 on page 23.)

★ (a) **Claim.** If A and B are sets and $\mathscr{P}(A) \subseteq \mathscr{P}(B)$, then $A \subseteq B$.

"Proof." $x \in A \Rightarrow \{x\} \subseteq A$
$\Rightarrow \{x\} \in \mathscr{P}(A)$
$\Rightarrow \{x\} \in \mathscr{P}(B)$
$\Rightarrow \{x\} \subseteq B$
$\Rightarrow x \in B.$

Therefore, $x \in A \Rightarrow x \in B$. Thus $A \subseteq B$. ■

(b) **Claim.** If A, B, and C are sets, and $A \subseteq B$, and $B \subseteq C$, then $A \subseteq C$.

"Proof." If $x \in C$, then, since $B \subseteq C$, $x \in B$. Since $A \subseteq B$ and $x \in B$, it follows that $x \in A$. Thus $x \in C$ implies $x \in A$. Therefore, $A \subseteq C$. ■

★ (c) **Claim.** If A, B, and C are sets, and $A \subseteq B$, and $B \subseteq C$, then $A \subseteq C$.

"Proof." Suppose x is any object. If $x \in A$, then $x \in B$, since $A \subseteq B$. If $x \in B$, then $x \in C$, since $B \subseteq C$. Therefore, $x \in C$. Therefore, $A \subseteq C$. ■

(d) **Claim.** If $X = \{x \in \mathbf{N} : x^2 < 14\}$ and $Y = \{1, 2, 3\}$, then $X = Y$.

"Proof." Since $1^2 = 1 < 14$, $2^2 = 4 < 14$, and $3^2 = 9 < 14$, $X = Y$. ■

SECTION 2.2. SET OPERATIONS

Given two sets A and B, there are several ways to combine them to produce a third set.

> **Definition.** Let A and B be sets. The **union** of A and B is defined by $A \cup B = \{x : x \in A \text{ or } x \in B\}$. The **intersection** of A and B is defined by $A \cap B = \{x : x \in A \text{ and } x \in B\}$. The **difference** of A and B is defined by $A - B = \{x : x \in A \text{ and } x \notin B\}$.

$A \cup B$ can be thought of as a new set formed from A and B by choosing as elements the objects that appear in at least one of A or B; $A \cap B$ consists of objects that appear in both; $A - B$ contains those elements of A that are not in B. See figures 2.1, 2.2, and 2.3. In case $A \cap B = \varnothing$, we say A and B are **disjoint**.

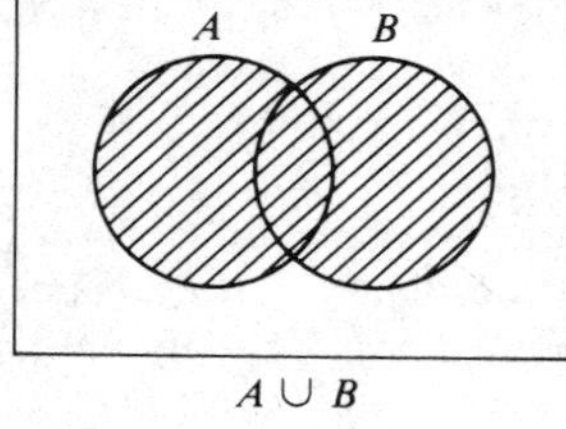

$A \cup B$

Figure 2.1

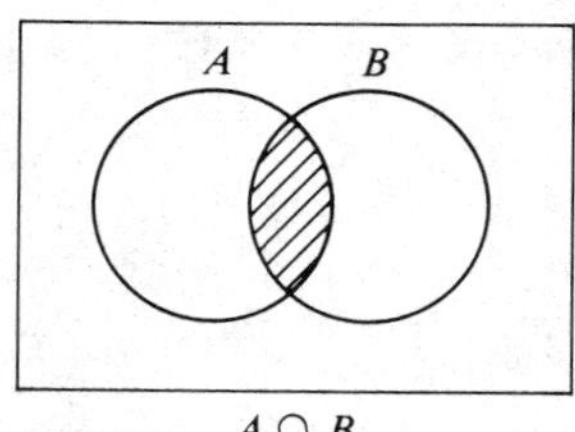

$A \cap B$

Figure 2.2

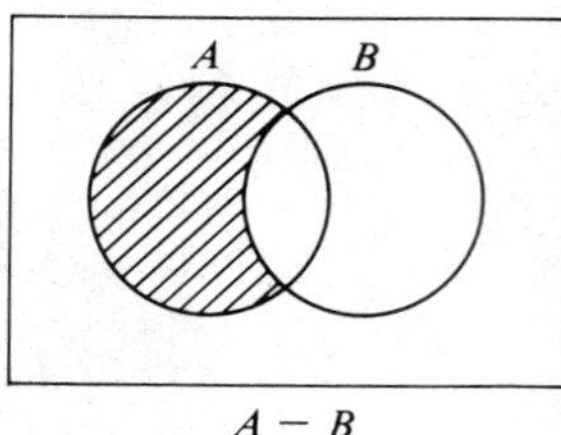

$A - B$

Figure 2.3

Example. For $A = \{1, 2, 4, 5, 7\}$ and $B = \{1, 3, 5, 9\}$,

$$A \cup B = \{1, 2, 3, 4, 5, 7, 9\}$$
$$A \cap B = \{1, 5\}$$
$$A - B = \{2, 4, 7\}$$

and

$$B - A = \{3, 9\}.$$

Example. For $C = [1, 4)$ and $D = (2, 6]$, $C \cup D = [1, 6]$, $C \cap D = (2, 4)$, and $C - D = [1, 2]$. The sets $[1, 2]$ and $(2, 4)$ are disjoint.

The set operations obey certain rules that at times allow us to simplify our work, in the same manner that the distributive law for real numbers can be used to simplify $12(2 + \frac{1}{3} + \frac{1}{4} + \frac{1}{6}) = 33$. Several such laws have been gathered together in the next theorem. Those not proved are left as exercises (see exercise 6).

Theorem 2.7. Let A, B, and C be sets. Then:

(a) $A \subseteq A \cup B$
(b) $A \cap B \subseteq A$
(c) $A \cap \varnothing = \varnothing$
(d) $A \cup \varnothing = A$
(e) $A \cap A = A$
(f) $A \cup A = A$
(g) $A \cup B = B \cup A$
(h) $A \cap B = B \cap A$
(i) $A - \varnothing = A$
(j) $\varnothing - A = \varnothing$
(k) $A \cup (B \cup C) = (A \cup B) \cup C$
(l) $A \cap (B \cap C) = (A \cap B) \cap C$
(m) $A \cap (B \cup C) = (A \cap B) \cup (A \cap C)$
(n) $A \cup (B \cap C) = (A \cup B) \cap (A \cup C)$
(o) $A \subseteq B$ iff $A \cup B = B$
(p) $A \subseteq B$ iff $A \cap B = A$
(q) If $A \subseteq B$, then $A \cup C \subseteq B \cup C$
(r) If $A \subseteq B$, then $A \cap C \subseteq B \cap C$

Proof. (b) We must show that, if $x \in A \cap B$, then $x \in A$. Suppose $x \in A \cap B$. Then $x \in A$ and $x \in B$. Therefore, ⟨*because* $P \wedge Q \Rightarrow P$,⟩ $x \in A$.

(f) We must show that $x \in A \cup A$ iff $x \in A$. By the definition of union, $x \in A \cup A$ iff $x \in A$ or $x \in A$. This is equivalent to $x \in A$. Therefore, $A \cup A = A$.

(h) ⟨*This iff proof uses the definition of intersection and the equivalence of* $P \wedge Q$ *and* $Q \wedge P$.⟩

$$\begin{aligned} x \in A \cap B &\text{ iff } x \in A \text{ and } x \in B \\ &\text{ iff } x \in B \text{ and } x \in A \\ &\text{ iff } x \in B \cap A. \end{aligned}$$

(m) ⟨*As you read this proof watch for the steps in which the definitions of union and intersection are used (two for each). Watch also for the use of the equivalence from Theorem 1.2 (f).*⟩

$$\begin{aligned} x \in A \cap (B \cup C) &\text{ iff } x \in A \text{ and } x \in B \cup C \\ &\text{ iff } x \in A \text{ and } (x \in B \text{ or } x \in C) \\ &\text{ iff } (x \in A \text{ and } x \in B) \text{ or } (x \in A \text{ and } x \in C) \\ &\text{ iff } x \in A \cap B \text{ or } x \in A \cap C \\ &\text{ iff } x \in (A \cap B) \cup (A \cap C). \end{aligned}$$

Therefore, $A \cap (B \cup C) = (A \cap B) \cup (A \cap C)$.

(p) ⟨*The statement* $A \subseteq B$ *iff* $A \cap B = A$ *requires separate proofs for each implication. We make use of earlier parts of this theorem.*⟩ First, assume that $A \subseteq B$. We must show that $A \cap B = A$. If $x \in A$, then from the hypothesis $A \subseteq B$, we have $x \in B$. Therefore, $x \in A$ implies $x \in A$ and $x \in B$, so $x \in A \cap B$. This shows that $A \subseteq A \cap B$, which, combined with $A \cap B \subseteq A$ from part (b) of this theorem, gives $A \cap B = A$.

Second, assume that $A \cap B = A$. We must show that $A \subseteq B$. By parts (b) and (h) of this theorem, we have $B \cap A \subseteq B$ and $B \cap A = A \cap B$. Therefore, $A \cap B \subseteq B$. By hypothesis, $A \cap B = A$, so $A \subseteq B$. ■

Recall that the universe of discourse is a collection of objects understood from the context or specified at the outset of a discussion and that all objects under consideration must belong to the universe.

Definition. If U is the universe and $B \subseteq U$, then we define the **complement** of B to be the set $\overline{B} = U - B$.

Thus $\overline{B}$ is to be interpreted as all those elements of the universe that are not in B. See figure 2.4.

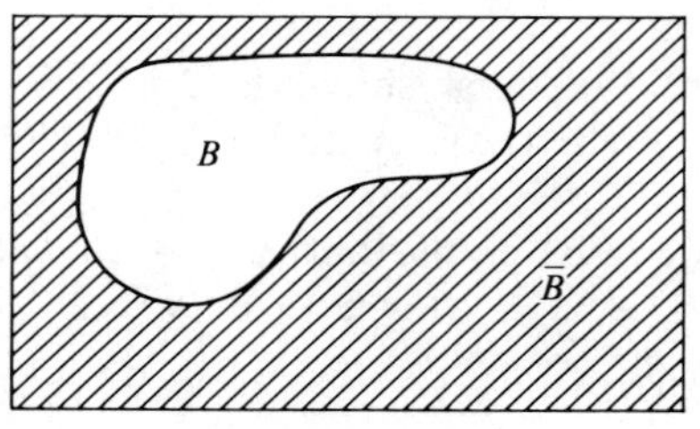

Figure 2.4

For the universe $\mathbf{R}$, if $B = (0, \infty)$ then $\bar{B} = (-\infty, 0]$. If $A = \{x: x \text{ is rational}\}$, then $\bar{A} = \{x: x \text{ is irrational}\}$.

The complement is nothing more than a short way to write set difference in the case where the first set is the universe. The complement operation obeys several rules.

Theorem 2.8. Let U be the universe, and let A and B be subsets of U. Then,

(a) $\bar{\bar{A}} = A$
(b) $A \cup \bar{A} = U$
(c) $A \cap \bar{A} = \varnothing$
(d) $A - B = A \cap \bar{B}$
(e) $A \subseteq B$ iff $\bar{B} \subseteq \bar{A}$
(f) $\overline{A \cup B} = \bar{A} \cap \bar{B}$ } De Morgan's Laws
(g) $\overline{A \cap B} = \bar{A} \cup \bar{B}$ }
(h) $A \cap B = \varnothing$ iff $A \subseteq \bar{B}$

Proof. (a) By definition of the complement, $x \in \bar{\bar{A}}$ iff $x \notin \bar{A}$ iff $x \in A$. Therefore $\bar{\bar{A}} = A$.

(e) ⟨*For this part of the theorem we give two separate proofs. The first proof has two parts, and its first part is used to prove its second part.*⟩

First proof of (e). We prove that if $A \subseteq B$, then $\bar{B} \subseteq \bar{A}$. Assume that $A \subseteq B$. Let $x \in \bar{B}$. Then $x \notin B$. Since $A \subseteq B$ and $x \notin B$, we have $x \notin A$. ⟨ *This is the contrapositive of* $A \subseteq B$.⟩ Thus $x \in \bar{A}$. Therefore, $\bar{B} \subseteq \bar{A}$. For the second part of this proof, we show that $\bar{B} \subseteq \bar{A}$ implies $A \subseteq B$. Assume that $\bar{B} \subseteq \bar{A}$. Then, by the first part of this proof, $\bar{\bar{A}} \subseteq \bar{\bar{B}}$. Therefore, using part (a), $A \subseteq B$. Combining the two parts of this proof, $A \subseteq B$ iff $\bar{B} \subseteq \bar{A}$.

Second proof of (e). ⟨*This proof makes use of the fact that a conditional sentence is equivalent to its contrapositive.*⟩

$$\begin{aligned} A \subseteq B &\text{ iff } (\forall x)(x \in A \Rightarrow x \in B) \\ &\text{ iff } (\forall x)(x \notin B \Rightarrow x \notin A) \\ &\text{ iff } (\forall x)(x \in \bar{B} \Rightarrow x \in \bar{A}) \\ &\text{ iff } \bar{B} \subseteq \bar{A}. \end{aligned}$$

(f) The object x is a member of $(\overline{A \cup B})$
iff x is not a member of $A \cup B$
iff it is not the case that $x \in A$ or $x \in B$
iff $x \notin A$ and $x \notin B$ ⟨*See Theorem 1.2* (c).⟩
iff $x \in \bar{A}$ and $x \in \bar{B}$
iff $x \in \bar{A} \cap \bar{B}$.

The proofs of the remaining parts are left as exercise 7. ■

Exercises 2.2

1. Let $A = \{1, 3, 5, 7, 9\}$, $B = \{0, 2, 4, 6, 8\}$, $C = \{1, 2, 4, 5, 7, 8\}$, and $D = \{1, 2, 3, 5, 6, 7, 8, 9, 10\}$. Find

★ (a) $A \cup B$ (b) $A \cap B$ ★ (c) $A - B$

(d) $A - (B - C)$ ★ (e) $(A - B) - C$ (f) $A \cap (C \cap D)$
★ (g) $(A \cap C) \cap D$ (h) $A \cap (B \cup C)$ ★ (i) $(A \cap B) \cup (A \cap C)$
(j) $(A \cup B) - (C \cap D)$

2. Let U be the set of all integers. Let E, D, P, and N be the sets of all even, odd, positive, and negative integers, respectively. Find
★ (a) $E - P$ (b) $P - E$ ★ (c) $D - E$
(d) $P - N$ ★ (e) $\bar{P}$ (f) $\bar{N}$
★ (g) $\bar{E}$ (h) $\bar{D}$ ★ (i) $E - N$
(j) $U - P$ ★ (k) $\bar{U}$ (l) $\bar{\varnothing}$

3. Let $U = \{1, 2, 3\}$ be the universe for the sets $A = \{1, 2\}$ and $B = \{2, 3\}$. Find
★ (a) $\mathscr{P}(A)$ (b) $\mathscr{P}(\bar{A})$
(c) $\mathscr{P}(A) \cap \mathscr{P}(B)$ ★ (d) $\mathscr{P}(\bar{A}) \cup \mathscr{P}(\bar{B})$
(e) $\mathscr{P}(A) - \mathscr{P}(B)$ ★ (f) $\mathscr{P}(\bar{A}) - \mathscr{P}(\bar{B})$

★ 4. Let A, B, C, and D be as in exercise 1. What pairs of sets are disjoint?
5. Let U, E, D, P, and N be as in exercise 2. What pairs of sets are disjoint?
6. Prove the remaining parts of Theorem 2.7.
7. Prove the remaining parts of Theorem 2.8.
★ 8. Let A and B be sets. Prove that $\mathscr{P}(A \cap B) = \mathscr{P}(A) \cap \mathscr{P}(B)$.
9. Let A, B, and C be sets. Prove that $(A - B) - C = (A - C) - (B - C)$.
★ 10. Show that there are no sets A and B such that $\mathscr{P}(A - B) = \mathscr{P}(A) - \mathscr{P}(B)$.
11. Let A and B be sets. Prove that $\mathscr{P}(A) \cup \mathscr{P}(B) \subseteq \mathscr{P}(A \cup B)$.
★ 12. Prove that if A and B are disjoint sets, $C \subseteq A$ and $D \subseteq B$, then C and D are disjoint.
13. Let A, B, and C be sets. Prove $C \subseteq A \cap B$ iff $C \subseteq A$ and $C \subseteq B$.
14. Provide examples of sets to show that each of the following is false.
★ (a) If $A \cup C \subseteq B \cup C$, then $A \subseteq B$. (b) If $A \cap C \subseteq B \cap C$, then $A \subseteq B$.
★ (c) $\mathscr{P}(A \cup B) = \mathscr{P}(A) \cup \mathscr{P}(B)$. (d) $\mathscr{P}(A) - \mathscr{P}(B) \subseteq \mathscr{P}(A - B)$.
★ (e) $A - (B - C) = (A - B) - (A - C)$.
15. Define the symmetric difference operation Δ on sets by: $A \Delta B = (A - B) \cup (B - A)$. Prove that:
(a) $A \Delta B = B \Delta A$.
(b) $A \Delta A = \varnothing$.
(c) $A \Delta \varnothing = A$.
(d) Show that $A \Delta B = (A \cup B) - (A \cap B)$.
16. **Proofs to Grade.**
★ (a) **Claim.** $A \subseteq B \Leftrightarrow A \cap B = A$.
"Proof." Assume that $A \subseteq B$. Let $x \in A \cap B$. Then $x \in A$ and $x \in B$, so $A \cap B = A$. Now, assume that $A \cap B = A$. Let $x \in A$. Then, $x \in A \cap B$, since $A = A \cap B$; and therefore, $x \in B$. This shows that $x \in A$ implies $x \in B$, and so $A \subseteq B$. ■
★ (b) **Claim.** $A \cap \varnothing = A$.
"Proof." We know that $x \in A \cap \varnothing$ iff $x \in A$ and $x \in \varnothing$. Since $x \in \varnothing$ is false, $x \in A$ and $x \in \varnothing$ iff $x \in A$. Therefore, $x \in A \cap \varnothing$ iff $x \in A$; that is, $A \cap \varnothing = A$. ■
(c) **Claim.** $\mathscr{P}(A - B) - \{\varnothing\} \subseteq \mathscr{P}(A) - \mathscr{P}(B)$.
"Proof." Let $x \in \mathscr{P}(A - B) - \{\varnothing\}$. Then $x \in \mathscr{P}(A) - \mathscr{P}(B)$. Therefore, $\mathscr{P}(A - B) \subseteq \mathscr{P}(A) - \mathscr{P}(B)$. ■
★ (d) **Claim.** $A \Delta A = \varnothing$. (Exercise 15 (b).)
"Proof." By part (d) of exercise 15, $A \Delta A = (A \cup A) - (A \cap A)$. By Theorem 2.7, parts (e) and (f), $(A \cup A) - (A \cap A) = A - A$. But $x \in A - A$ iff both $x \in A$ and $x \notin A$ iff $x \in \varnothing$. ⟨*Each is false.*⟩ Therefore, $A \Delta A = \varnothing$. ■

(e) **Claim.** $C \subseteq A \cup B$ iff $C \subseteq A$ or $C \subseteq B$.
"Proof." If $x \in C$, then $x \in A \cup B$
$\Leftrightarrow$ If $x \in C$, then $x \in A$ or $x \in B$.
$\Leftrightarrow$ If $x \in C$, then $x \in A$ or, if $x \in C$, then $x \in B$.
$\Leftrightarrow$ $C \subseteq A$ or $C \subseteq B$. ■

★ (f) **Claim.** If $A \subseteq B$, then $A \cup B = B$.
"Proof." Let $A \subseteq B$. Then A and B are related as in this figure.

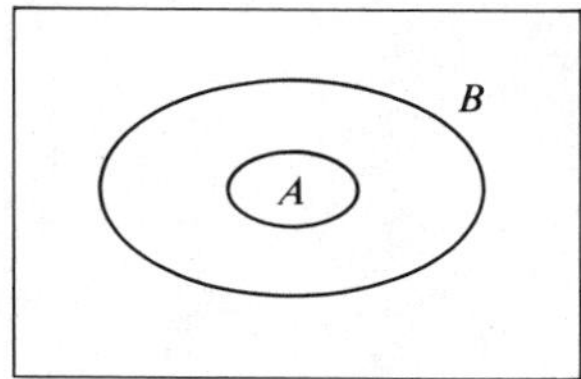

Since $A \cup B$ is the set of elements in either of the sets A or B, $A \cup B$ is the shaded area in

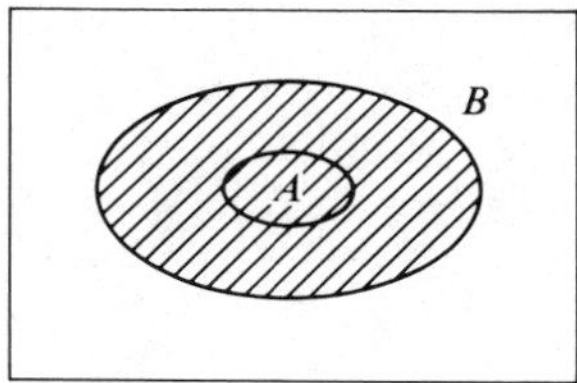

Since this is B, $A \cup B = B$. ■

SECTION 2.3. EXTENDED SET OPERATIONS AND INDEXED FAMILIES OF SETS

A set of sets is often called a **family** of sets. In this section, we extend the definition of union and intersection to families of sets, so that the union and intersection of two sets will be a special case. If $\mathscr{A}$ is a family of sets, the intersection over $\mathscr{A}$ will be the set of all elements common to all sets in $\mathscr{A}$, while the union over $\mathscr{A}$ will consist of those objects appearing in at least one of the sets in $\mathscr{A}$. See figure 2.5.

Definition. Let $\mathscr{A}$ be a family of sets. The **union over** $\mathscr{A}$ is defined by $\bigcup_{A \in \mathscr{A}} A = \{x : (\exists A \in \mathscr{A})(x \in A)\}$ and the **intersection over** $\mathscr{A}$ is $\bigcap_{A \in \mathscr{A}} A = \{x : (\forall A \in \mathscr{A})(x \in A)\}$.

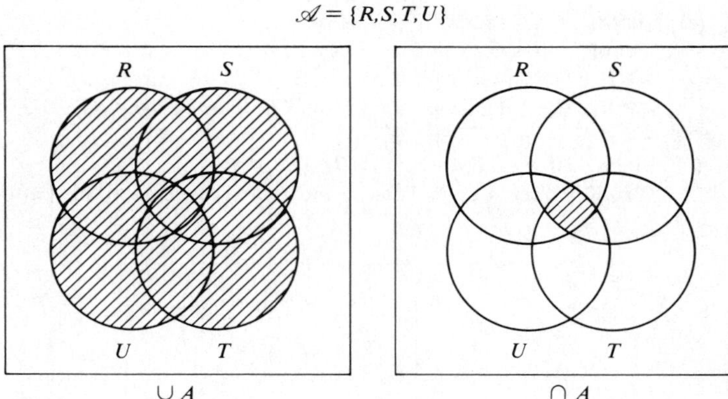

Figure 2.5

Let $\mathscr{A} = \{\{a, b, c\}, \{b, c, d\}, \{c, d, e, f\}\}$ be a family of three sets. Then $\bigcup_{A \in \mathscr{A}} A = \{a, b, c, d, e, f\}$, since each of the elements a, b, c, d, e, f appears in at least one of the three sets in $\mathscr{A}$. Also, $\bigcap_{A \in \mathscr{A}} A = \{c\}$, since c is the only object appearing in every set in $\mathscr{A}$.

Theorem 2.9. For every set B in a family $\mathscr{A}$ of sets, $\bigcap_{A \in \mathscr{A}} A \subseteq B$ and $B \subseteq \bigcup_{A \in \mathscr{A}} A$.

Proof. Let $\mathscr{A}$ be a family of sets and $B \in \mathscr{A}$. If $x \in \bigcap_{A \in \mathscr{A}} A$, then $x \in A$ for every $A \in \mathscr{A}$. ⟨*Notice that the set A in the last sentence is a dummy symbol. It stands for any set in the family. The set B is in the family.*⟩ In particular, $x \in B$. Therefore, $\bigcap_{A \in \mathscr{A}} A \subseteq B$. The proof that B is a subset of the union over $\mathscr{A}$ is left as an exercise (exercise 12). ■

Let $\mathscr{A}$ be the set of all subsets of the real line of the form $(-a, a)$ for some positive $a \in \mathbf{R}$. For example, $(-1, 1) \in \mathscr{A}$ and $(-\sqrt{2}, \sqrt{2}) \in \mathscr{A}$. Then, $\bigcup_{A \in \mathscr{A}} A = \mathbf{R}$ and $\bigcap_{A \in \mathscr{A}} A = \{0\}$. See figure 2.6.

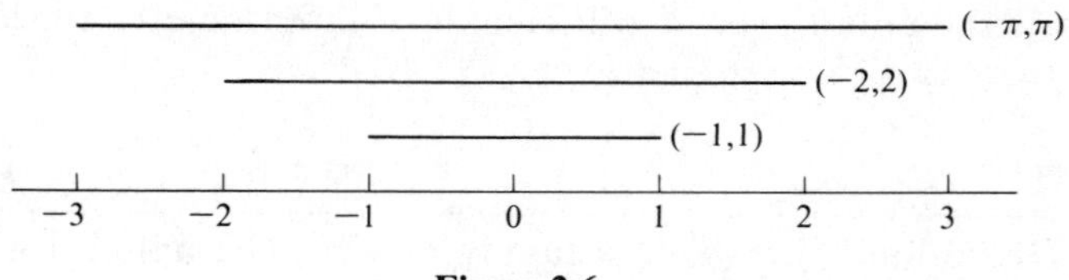

Figure 2.6

If $\mathscr{B}$ is the family of all sets of the form $(-5, n]$, where n is a positive even integer, then $\bigcup_{A \in \mathscr{B}} A = (-5, +\infty)$ and $\bigcap_{A \in \mathscr{B}} A = (-5, 2]$. Some members of the family $\mathscr{B}$ are $A_2 = (-5, 2]$, $A_4 = (-5, 4]$, and $A_{28} = (-5, 28]$. It is often possible, as it is in this case, to associate a kind of identifying tag, or index, with each set in a family. For the family $\mathscr{B}$ and set $(-5, 2]$, the index used above is 2, and 28 is the index for the set $(-5, 28]$. With this method of indexing, the set of indices for $\mathscr{B}$ is the set of all positive even integers.

Definition. Let Δ be a nonempty set. Suppose for each $\alpha \in \Delta$, there is a corresponding set A_α. Then the family of sets $\mathscr{A} = \{A_\alpha: \alpha \in \Delta\}$ is an **indexed family of sets.** Each $\alpha \in \Delta$ is called an **index** and Δ is called an **indexing set.**

There is no real difference between a family of sets and an indexed family. Every family of sets could be indexed by finding a large enough set of indices to label each set in the family. An indexing set may be finite or infinite, as well as ordered (like the integers) or unordered (like the complex numbers).

Example. For the sets $A_1 = \{1, 2, 4, 5\}$, $A_2 = \{2, 3, 5, 6\}$, and $A_3 = \{3, 4, 5, 6\}$, the index set has been chosen to be $\Delta = \{1, 2, 3\}$. The family $\mathscr{A}$ indexed by Δ is $\mathscr{A} = \{A_1, A_2, A_3\} = \{A_i: i \in \Delta\}$. The family $\mathscr{A}$ could be indexed by another set. For instance, if $\Gamma = \{10, 21, \pi\}$, and $A_{10} = \{1, 2, 4, 5\}$, $A_{21} = \{2, 3, 5, 6\}$, $A_\pi = \{3, 4, 5, 6\}$, then $\{A_i: i \in \Delta\} = \{A_i: i \in \Gamma\}$.

Unions and intersections over indexed families of sets are denoted as follows. Note the relationship between $\cup$ and $\exists$, and that between $\cap$ and $\forall$. This occurs because an element of the union is a member of **at least one** set in the family, while an element of the intersection is in **every** set in the family.

Definition. If $\mathscr{A} = \{A_\alpha: \alpha \in \Delta\}$ is an indexed family of sets, then

$$\bigcup_{\alpha \in \Delta} A_\alpha = \bigcup_{A \in \mathscr{A}} A = \{x: (\exists \alpha \in \Delta)(x \in A_\alpha)\}$$

and

$$\bigcap_{\alpha \in \Delta} A_\alpha = \bigcap_{A \in \mathscr{A}} A = \{x: (\forall \alpha \in \Delta)(x \in A_\alpha)\}.$$

Example. In the previous example, where $A_1 = \{1, 2, 4, 5\}$, $A_2 = \{2, 3, 5, 6\}$, $A_3 = \{3, 4, 5, 6\}$, and $\mathscr{A} = \{A_i: i \in \Delta\}$, $\bigcap_{i \in \Delta} A_i = \{5\}$ and $\bigcup_{i \in \Delta} A_i = \{1, 2, 3, 4, 5, 6\}$.

Example. Let Δ be the set of positive reals. For each $a \in \Delta$, let $H_a = (-a, a)$. Let $\mathscr{A} = \{H_a: a \in \Delta\}$. Then, as we have seen before, $\bigcap_{a \in \Delta} H_a = \{0\}$ and $\bigcup_{a \in \Delta} H_a = \mathbf{R}$.

Example. For each real number x, define $B_x = [x^2, x^2 + 1]$. Then $B_{-1/2} = [\frac{1}{4}, \frac{5}{4}]$, $B_0 = [0, 1]$, and $B_{10} = [100, 101]$. In this example we have different indices representing the same set. For example, $B_{-2} = B_2 = [4, 5]$. Here the index set is $\mathbf{R}$, $\bigcap_{x \in \mathbf{R}} B_x = \varnothing$ and $\bigcup_{x \in \mathbf{R}} B_x = [0, +\infty)$.

Example. Let A be any nonempty set. For each $a \in A$, let $X_a = \{a\}$. Then, $\bigcup_{a \in A} X_a = A$. If A has more than one element, then $\bigcap_{a \in A} X_a = \varnothing$.

Families of sets, whether indexed or not, obey a form of De Morgan's laws stated for two sets in Theorem 2.8. The next theorem gives a statement of De Morgan's laws for indexed families, and also restates Theorem 2.9 for indexed families.

Theorem 2.10. Let $\mathscr{A} = \{A_\alpha : \alpha \in \Delta\}$ be an indexed collection of sets. Then

(a) $\bigcap_{\alpha \in \Delta} A_\alpha \subseteq A_\beta$ for each $\beta \in \Delta$.

(b) $A_\beta \subseteq \bigcup_{\alpha \in \Delta} A_\alpha$ for each $\beta \in \Delta$.

(c) $\overline{\bigcap_{\alpha \in \Delta} A_\alpha} = \bigcup_{\alpha \in \Delta} \overline{A_\alpha}$.

(d) $\overline{\bigcup_{\alpha \in \Delta} A_\alpha} = \bigcap_{\alpha \in \Delta} \overline{A_\alpha}$.

Proof. The proofs of parts (a) and (b) are similar to those for Theorem 2.9.

(c) $x \in \overline{\bigcap_{\alpha \in \Delta} A_\alpha}$ iff $x \notin \bigcap_{\alpha \in \Delta} A_\alpha$

iff it is not the case that for every $\alpha \in \Delta$, $x \in A_\alpha$.
iff for some $\beta \in \Delta$, $x \notin A_\beta$.
iff for some $\beta \in \Delta$, $x \in \overline{A}_\beta$.
iff $x \in \bigcup_{\alpha \in \Delta} \overline{A_\alpha}$.

Therefore, $\overline{\bigcap_{\alpha \in \Delta} A_\alpha} = \bigcup_{\alpha \in \Delta} \overline{A_\alpha}$.

(d) ⟨ *One proof of part (d) is very similar to that given for part (c). However, since part (c) has been proved, it is permissible to use it, as follows.* ⟩

$$\begin{aligned} \overline{\bigcup_{\alpha \in \Delta} A_\alpha} &= \overline{\bigcup_{\alpha \in \Delta} \overline{\overline{A}}_\alpha} && \langle A_\alpha = \overline{\overline{A}} \rangle \\ &= \overline{\overline{\bigcap_{\alpha \in \Delta} \overline{A}_\alpha}} && \langle \textit{by (c)} \rangle \\ &= \bigcap_{\alpha \in \Delta} \overline{A_\alpha} && \langle \overline{\overline{A}} = A \rangle. \quad \blacksquare \end{aligned}$$

Definition. The family $\mathscr{A} = \{A_\alpha : \alpha \in \Delta\}$ of sets is **pairwise disjoint** iff for all α and β in Δ, if $A_\alpha \neq A_\beta$, then $A_\alpha \cap A_\beta = \varnothing$.

Two questions are commonly asked about this definition. The first question is why we bother with such a definition when we could more easily talk about a disjoint family satisfying $\bigcap_{\alpha \in \Delta} A_\alpha = \varnothing$. The answer is that the two ways of talking about disjointness of families of sets are not the same, and the definition given is by far the more interesting. A pairwise disjoint family with more than one set always satisfies the condition that the intersection over the whole family is empty. One reason why the pairwise disjoint idea is important will appear in the study of partitions (chapter 3, section 3). The fact that the two ideas of disjointness are not the same can be seen from the following examples.

Let $\mathscr{C} = \{A_1, A_2, A_3\}$, where $A_1 = \{1, 2\}$, $A_2 = \{2, 3\}$, and $A_3 = \{4, 5\}$. Then $\bigcap_{A \in \mathscr{C}} A = \varnothing$, but the family is not pairwise disjoint because $A_1 \cap A_2 \neq \varnothing$.

For the family $\mathscr{A} = \{A_n : n \in \mathbf{N}\}$ where, for each n, $A_n = (n, \infty)$, the family is not pairwise disjoint, but $\bigcap_{n \in \mathbf{N}} A_n = \varnothing$. In this case not only is there some pairs of sets that are not disjoint, but for every pair A_n, A_m, with, say $n < m$, the intersection $A_n \cap A_m = (m, \infty) = A_m$.

The second common question about the definition of pairwise disjoint has to do with the need for including the phrase ". . . whenever $A_\alpha \neq A_\beta$" in the definition. This is necessary because the family $\{A_1, A_2, A_3, A_4\}$ with $A_1 = \{6, 7\}$, $A_2 = \{5, 9\}$, $A_3 = \{6, 7\}$, and $A_4 = \{2, 8\}$ is the same family as $\{A_1, A_2, A_4\}$ and is pairwise disjoint. We do not require that two sets with different indices be different.

Exercises 2.3

★ 1. Let $\mathscr{A} = \{\{1, 2, 3, 4, 5\}, \{2, 3, 4, 5, 6\}, \{3, 4, 5, 6, 7\}, \{4, 5, 6, 7, 8\}\}$. Find $\bigcup_{A \in \mathscr{A}} A$ and $\bigcap_{A \in \mathscr{A}} A$.

2. Let $\mathscr{A} = \{\{1, 3, 5\}, \{2, 4, 6\}, \{7, 9, 11, 13\}, \{8, 10, 12\}\}$. Find $\bigcup_{A \in \mathscr{A}} A$ and $\bigcap_{A \in \mathscr{A}} A$.

★ 3. For each natural number n, let $A_n = \{1, 2, 3, \ldots, n\}$, and let $\mathscr{A} = \{A_n : n \in \mathbf{N}\}$. Find $\bigcup_{i \in \mathbf{N}} A_i$ and $\bigcap_{i \in \mathbf{N}} A_i$.

4. For each natural number n, let $B_n = \mathbf{N} - \{1, 2, 3, \ldots, n\}$, and let $\mathscr{B} = \{B_n : n \in \mathbf{N}\}$. Find $\bigcup_{B \in \mathscr{B}} B$ and $\bigcap_{B \in \mathscr{B}} B$.

★ 5. Let $\mathscr{A}$ be the set of all sets of integers that contain 10. Find $\bigcup_{A \in \mathscr{A}} A$ and $\bigcap_{A \in \mathscr{A}} A$.

6. Let $A_1 = \{1\}$, $A_2 = \{2, 3\}$, $A_3 = \{3, 4, 5\}, \ldots, A_{10} = \{10, 11, \ldots, 19\}$; and let $\mathscr{A} = \{A_n : n \in \{1, 2, 3, \ldots, 10\}\}$. Find $\bigcup_{A \in \mathscr{A}} A$ and $\bigcap_{A \in \mathscr{A}} A$.

★ 7. For each natural number, let $A_n = (0, 1/n)$, and let $\mathscr{A} = \{A_n : n \in \mathbf{N}\}$. Find $\bigcup_{n \in \mathbf{N}} A_n$ and $\bigcap_{n \in \mathbf{N}} A_n$.

8. Let $\mathbf{R}^+ = (0, \infty)$. For each positive real number r, let $A_r = [-\pi, r)$, and let $\mathscr{A} = \{A_r : r \in \mathbf{R}^+\}$. Find $\bigcup_{r \in \mathbf{R}^+} A_r$ and $\bigcap_{r \in \mathbf{R}^+} A_r$.

★ 9. For each real number r, let $A_r = [\,|r|, 2|r| + 1]$, and let $\mathscr{A} = \{A_r : r \in \mathbf{R}\}$. Find $\bigcup_{r \in \mathbf{R}} A_r$ and $\bigcap_{r \in \mathbf{R}} A_r$.

10. For each $n \in \mathbf{N}$, let $M_n = \{\ldots, -3n, -2n, -n, 0, n, 2n, 3n, \ldots\}$, and let $\mathscr{M} = \{M_n : n \in \mathbf{N}\}$. Find $\bigcup_{n \in \mathbf{N}} M_n$ and $\bigcap_{n \in \mathbf{N}} M_n$.

☆ 11. Which families in exercises 1 through 10 are pairwise disjoint?

12. Prove the remaining part of Theorem 2.9. That is, prove that if $\mathscr{A}$ is a family of sets and $B \in \mathscr{A}$, then $B \subseteq \bigcup_{A \in \mathscr{A}} A$.

☆ 13. Prove parts (a) and (b) of Theorem 2.10.

14. Give a direct proof of part (d) of Theorem 2.10 that does not use part (c).

15. Let $\mathscr{A} = \{A_\alpha : \alpha \in \Delta\}$ be an arbitrary family of sets and let B be a set. Prove that

★ (a) $B \cap \bigcup_{\alpha \in \Delta} A_\alpha = \bigcup_{\alpha \in \Delta} (B \cap A_\alpha)$.

(b) $B \cup \bigcap_{\alpha \in \Delta} A_\alpha = \bigcap_{\alpha \in \Delta} (B \cup A_\alpha)$.

16. Let $\mathscr{A} = \{A_\alpha : \alpha \in \Delta\}$ and $\mathscr{B} = \{B_\beta : \beta \in \Gamma\}$. Use exercise 15 to write

★ (a) $(\bigcup_{\alpha \in \Delta} A_\alpha) \cap (\bigcup_{\beta \in \Gamma} B_\beta)$ as a union of intersections.

(b) $(\bigcap_{\alpha \in \Delta} A_\alpha) \cup (\bigcap_{\beta \in \Gamma} B_\beta)$ as an intersection of unions.

17. Let $\mathscr{A}$ be a family of sets and suppose $\varnothing \in \mathscr{A}$. Prove that $\bigcap_{A \in \mathscr{A}} A = \varnothing$.

18. If $\mathscr{A} = \{A_\alpha : \alpha \in \Delta\}$ is a family of sets and if $\Gamma \subseteq \Delta$, prove that

★ (a) $\bigcup_{\alpha \in \Gamma} A_\alpha \subseteq \bigcup_{\alpha \in \Delta} A_\alpha$.

(b) $\bigcap_{\alpha \in \Delta} A_\alpha \subseteq \bigcap_{\alpha \in \Gamma} A_\alpha$.

19. Let $\mathscr{A}$ be a family of sets and suppose $B \subseteq A$ for every $A \in \mathscr{A}$. Prove that $B \subseteq \bigcap_{A \in \mathscr{A}} A$.

20. Let $\mathscr{A}$ be a family of sets and suppose that $A \subseteq B$ for every $A \in \mathscr{A}$. Prove that $\bigcup_{A \in \mathscr{A}} A \subseteq B$.

21. Prove that, if $\mathscr{A} = \{A_\alpha : \alpha \in \Delta\}$ is pairwise disjoint and Δ contains more than one element, then $\bigcap_{\alpha \in \Delta} A_\alpha = \varnothing$.

★ 22. Let $\mathscr{A} = \{A_\alpha : \alpha \in \Delta\}$ be a family of sets. By Theorem 2.10 we know that $\bigcap_{\alpha \in \Delta} A_\alpha \subseteq A_\beta$ for every $\beta \in \Delta$. Prove that $\bigcap_{\alpha \in \Delta} A_\alpha$ is the largest such set by proving that if B is any other set such that $B \subseteq A_\beta$ for all $\beta \in \Delta$, then $B \subseteq \bigcap_{\alpha \in \Delta} A_\alpha$.

23. Let $\mathscr{A} = \{A_\alpha : \alpha \in \Delta\}$ be a family of sets. By Theorem 2.10 we know that $A_\beta \subseteq \bigcup_{\alpha \in \Delta} A_\alpha$ for every $\beta \in \Delta$. Prove that $\bigcup_{\alpha \in \Delta} A_\alpha$ is the smallest such set by proving that if B is any other set such that $A_\beta \subseteq B$ for all $\beta \in \Delta$, then $\bigcup_{\alpha \in \Delta} A_\alpha \subseteq B$.

★ 24. Give another example of an indexed collection of sets $\{A_\alpha : \alpha \in \Delta\}$ such that, for all α and $\beta \in \Delta$, $A_\alpha \cap A_\beta \neq \varnothing$, but $\bigcap_{\alpha \in \Delta} A_\alpha = \varnothing$.

25. Prove that every family $\mathscr{A}$ of sets can be written as an indexed family. (*Hint:* Can a set of sets be used to index itself?)

26. **Proofs to Grade.**

★ (a) **Claim.** For any indexed family $\{A_\alpha : \alpha \in \Delta\}$, $\bigcap_{\alpha \in \Delta} A_\alpha \subseteq \bigcup_{\alpha \in \Delta} A_\alpha$.

"Proof." Choose any $A_\beta \in \{A_\alpha : \alpha \in \Delta\}$. Then, $\bigcap_{\alpha \in \Delta} A_\alpha \subseteq A_\beta$ and $A_\beta \subseteq \bigcup_{\alpha \in \Delta} A_\alpha$. Therefore, by transitivity of set inclusion, $\bigcap_{\alpha \in \Delta} A_\alpha \subseteq \bigcup_{\alpha \in \Delta} A_\alpha$. ■

★ (b) **Claim.** If $A_\alpha \subseteq B$ for all $\alpha \in \Delta$, then $\bigcup_{\alpha \in \Delta} A_\alpha \subseteq B$.

"Proof." Let $x \in \bigcup_{\alpha \in \Delta} A_\alpha$. Then, since $A_\alpha \subseteq B$ for all $\alpha \in \Delta$, $x \in B$. Therefore, $\bigcup_{\alpha \in \Delta} A \subseteq B$. ■

(c) **Claim.** For any indexed family $\{A_\alpha: \alpha \in \Delta\}$, $\bigcup_{\alpha \in \Delta} A_\alpha \subseteq \bigcap_{\alpha \in \Delta} A_\alpha$.

"Proof." Let $x \in \bigcup_{\alpha \in \Delta} A_\alpha$. Then $x \in A_\alpha$ for every $\alpha \in \Delta$, which implies $x \in A_\alpha$ for at least one $\alpha \in \Delta$ since $\Delta \neq \varnothing$. Therefore, $x \in \bigcap_{\alpha \in \Delta} A_\alpha$. ■

(d) **Claim.** For any indexed family $\{A_\alpha: \alpha \in \Delta\}$, $\bigcap_{\alpha \in \Delta} A_\alpha \subseteq \bigcup_{\alpha \in \Delta} A_\alpha$.

"Proof." Assume $\bigcap_{\alpha \in \Delta} A_\alpha \not\subseteq \bigcup_{\alpha \in \Delta} A_\alpha$. Then, for some $x \in \bigcap_{\alpha \in \Delta} A_\alpha$, $x \notin \bigcup_{\alpha \in \Delta} A_\alpha$. Since $x \notin \bigcup_{\alpha \in \Delta} A_\alpha$, it is not the case that $x \in A_\alpha$ for some $\alpha \in \Delta$. Therefore, $x \notin A_\alpha$ for every $\alpha \in \Delta$. But since $x \in \bigcap_{\alpha \in \Delta} A_\alpha$, $x \in A_\alpha$ for every $\alpha \in \Delta$. This is a contradiction, so we conclude $\bigcap_{\alpha \in \Delta} A_\alpha \subseteq \bigcup_{\alpha \in \Delta} A_\alpha$. ■

SECTION 2.4. INDUCTION

The most fundamental number system is the system of natural numbers. The set $\mathbf{N} = \{1, 2, 3, 4, \ldots\}$ and its addition, multiplication, and properties of closure, associativity, commutativity, distributivity, and order properties are well known. These properties will be assumed as we focus on the induction property and proofs by induction.

Principle of Mathematical Induction (PMI). Let S be a subset of $\mathbf{N}$ with the properties

(i) $1 \in S$.
(ii) For all $n \in \mathbf{N}$, $n \in S$ implies $n + 1 \in S$.

Then $S = \mathbf{N}$.

A set of natural numbers with the property that $n \in S$ implies $n + 1 \in S$ is called an **inductive set.** The set $\{5, 6, 7, 8, \ldots\}$ is inductive but does not contain 1. The set $\{1, 3, 5, 7, 9, \ldots\}$ contains 1 but is not inductive because, for example, 7 is a member but 8 is not. Many sets of natural numbers have the inductive property or contain 1, but only one set contains 1 and is also inductive. That set is $\mathbf{N}$.

We shall see that there are several forms of induction, each useful in certain situations. The PMI in the form given is basic to our understanding of the natural numbers: the natural numbers consist of 1, then $1 + 1 = 2$, then $2 + 1 = 3, \ldots$, and so on. The PMI allows us to do two important things: first, to make inductive definitions; and second, to prove that some properties are shared by all natural numbers.

You have probably used mathematical induction to prove statements like the one in the following example.

Example. For every natural number n, $1 + 3 + 5 + \cdots + (2n - 1) = n^2$. The correctness of this statement can be verified for, say, $n = 6$, by adding $1 + 3 + 5 + 7 + 9 + 11 = 36$. A proof by induction that the property holds for all n proceeds as follows:

Proof. Let $S = \{n \in \mathbf{N} : 1 + 3 + 5 + \cdots + (2n - 1) = n^2\}$. ⟨*Our aim is to show that the statement is true for every natural number by showing that* $S = \mathbf{N}$.⟩

(i) $1 = 1^2$, so $1 \in S$.
(ii) Let n be a natural number such that $n \in S$.

Then $1 + 3 + 5 + \cdots + (2n - 1) = n^2$. ⟨*We have* not *assumed what is to be proved. We assume only that* some *n is in S in order to show that* $n + 1 \in S$ *follows.*⟩ Therefore,

$$1 + 3 + 5 + \cdots + (2n - 1) + (2(n + 1) - 1) = n^2 + 2(n + 1) - 1$$
⟨*Compare the statements for n and for n + 1.*⟩
$$= n^2 + 2n + 1 = (n + 1)^2.$$

This shows that $n + 1 \in S$.

By the PMI, $S = \mathbf{N}$. That is, $1 + 3 + 5 + \cdots + (2n - 1) = n^2$ for every natural number n. ■

You are familiar with the distributive law that says that for any numbers m, x_1, and x_2, $m(x_1 + x_2) = mx_1 + mx_2$. It is also true that no matter how many numbers are involved, the numbers $m(x_1 + x_2 + \cdots + x_n)$ and $mx_1 + mx_2 + \cdots + mx_n$ are equal. Although we would otherwise accept this as being obviously true, we will, as an example, give a proof of this generalization of the distributive law. Since $+$ is defined only for two numbers, we must be certain to make a sum $x_1 + x_2 + \cdots + x_n$ of n numbers meaningful by thinking of it as a sum of two numbers. We use the following inductive definition:

$$\text{For } n \geq 3,\ x_1 + x_2 + \cdots + x_n = (x_1 + x_2 + \cdots + x_{n-1}) + x_n.$$

This definition assures us that a sum of n numbers has been unambiguously defined for every natural number n. The sum of one number is understood to be the number; we know all about the sum of two numbers; $x_1 + x_2 + x_3$ is the sum of $x_1 + x_2$ and x_3; $x_1 + x_2 + x_3 + x_4$ is the sum of $x_1 + x_2 + x_3$ and x_4; and so on. We are now prepared to prove the generalized distributive law.

Example. For every list $x_1, x_2, \ldots, x_n$ of n numbers, and every number m, $m(x_1 + x_2 + \cdots + x_n) = mx_1 + mx_2 + \cdots + mx_n$.

Proof. Let S be the set of natural numbers with the desired property. That is, $S = \{n \in \mathbf{N}$: for every list $x_1, x_2, \ldots, x_n$ of n numbers,
$$m(x_1 + x_2 + \cdots + x_n) = mx_1 + \cdots + mx_n\}.$$

(i) $m(r_1) = mr_1$ for every list of 1 number, so $1 \in S$.
(ii) Suppose $n \in S$, and consider the situation for a list $x_1, x_2, \ldots, x_{n+1}$ of $n + 1$ numbers. Then

$m(x_1 + x_2 + \cdots + x_{n+1})$
$= m((x_1 + x_2 + \cdots + x_n) + x_{n+1})$ ⟨*by definition*⟩
$= m(x_1 + x_2 + \cdots + x_n) + mx_{n+1}$ ⟨*by the distributive law*⟩
$= (mx_1 + mx_2 + \cdots + mx_n) + mx_{n+1}$ ⟨*by the assumption that* $n \in S$⟩
$= mx_1 + mx_2 + \cdots + mx_n + mx_{n+1}$ ⟨*by definition*⟩.

Therefore, $n + 1 \in S$.

By the PMI, $S = \mathbf{N}$. That is, the generalized distributive law holds for the sum of n numbers, for every $n \in \mathbf{N}$. ∎

Most people, when they write induction proofs, do not define a set S of all numbers that satisfy the property in question. It is usual to show simply that 1 has the property and to show that for every natural number n, if n has the property, then $n + 1$ has the property. The assumption that n has the property is called the **hypothesis of induction.** Every good proof by induction will use its hypothesis of induction to show that $n + 1$ has the desired property.

Example. For all $n \in \mathbf{N}$, $n + 3 < 5n^2$.

Proof.

(i) $1 + 3 < 5 \cdot 1^2$, so the statement is true for 1.
(ii) Assume that for some n, $n + 3 < 5n^2$. Then
$(n + 1) + 3 = n + 3 + 1 < 5n^2 + 1$ ⟨*by the hypothesis of induction*⟩,
and $5n^2 + 1 < 5n^2 + 10n + 5 = 5(n + 1)^2$; so
$(n + 1) + 3 < 5(n + 1)^2$. Thus the property is true for $n + 1$.

By the PMI, $n + 3 < 5n^2$ for every natural number n. ∎

Example. It is easy to see that $x - y$ divides $x^2 - y^2$ and $x^3 - y^3$ because $x^2 - y^2 = (x - y)(x + y)$ and $x^3 - y^3 = (x - y)(x^2 + xy + y^2)$. This suggests the possibility that for every natural number n, $x - y$ divides $x^n - y^n$. We prove this by induction.

Proof.

(i) $x - y$ obviously divides $x^1 - y^1 = x - y$, so the statement holds for $n = 1$.
(ii) Assume that $x - y$ divides $x^n - y^n$ for some n. We must show that $x - y$ divides $x^{n+1} - y^{n+1}$. We write $x^{n+1} - y^{n+1} = xx^n - yy^n = xx^n - yx^n + yx^n - yy^n = (x - y)x^n + y(x^n - y^n)$. Now $x - y$ divides the first term because it divides $x - y$ and the second term because it divides $x^n - y^n$ ⟨*by the hypothesis of induction*⟩. Therefore, $x - y$ divides the sum; so $x - y$ divides $x^{n+1} - y^{n+1}$.

By the PMI, $x - y$ divides $x^n - y^n$ for every natural number n. ∎

As our last example of the use of mathematical induction, we show that every natural number n is either 1 or else is a successor—that is, $n = k + 1$ for some natural number k. The number 1 is not a successor, because 0 is not a natural number. We will use this theorem in our proof of the example on page 51.

Theorem 2.11. For every $n \in \mathbf{N}$, $n = 1$ or n is a successor.

Proof. Let $M = \{1\} \cup \{a \in \mathbf{N} : (\exists b \in \mathbf{N})(a = b + 1)\}$. Then $1 \in M$. Suppose $n \in M$. Then $n \in \mathbf{N}$ and $n + 1$ is a natural number that is a successor ⟨*of n*⟩. Therefore, $n + 1 \in M$. By the PMI, $M = \mathbf{N}$. ∎

In some cases using the PMI is not the most convenient form of inductive proof. To use the PMI successfully, one must be able to show that if n has a certain property, then $n + 1$ has the property. In some cases there is no apparent connection between the property for n and for $n + 1$. An alternate form of induction is needed.

Theorem 2.12. Principle of Complete Induction (PCI). Let S be a subset of $\mathbf{N}$ with the following property:

for all natural numbers, m, if $\{1, 2, \ldots, m - 1\} \subseteq S$, then $m \in S$.
Then $S = \mathbf{N}$.

Proof. Let S be a set such that, for all m, if $\{1, \ldots, m - 1\} \subseteq S$, then $m \in S$. We use the PMI to show $S = \mathbf{N}$.

(i) First, $1 \in S$, since every natural number less than 1 is in S. ⟨*There are none.*⟩
(ii) Assume now that $n \in S$. We must show that $n + 1 \in S$. We know that $2 \in S$, since all numbers less than 2 are in S, ⟨$1 \in S$,⟩ and it follows that $3 \in S$, and so on. We successively argue ⟨*in only finitely many steps*⟩ that each number less than n is in S. Therefore ⟨*recalling* $n \in S$⟩, all natural numbers less than $n + 1$ are in S, so by the condition on S, $n + 1 \in S$.

By the PMI, $S = \mathbf{N}$. ■

The PCI could have been accepted rather than the PMI as the fundamental property of $\mathbf{N}$, because the two are equivalent. Notice that the PCI does not require the separate step of showing that $1 \in S$. However, in order to prove that **for all** m, $\{1, 2, \ldots, m - 1\} \subseteq S$ implies $m \in S$, special considerations may be necessary when $m = 1$ or $m = 2$. This caution may apply to the PMI, too, when $m = 2$. (See exercise 12 (a).)

Both the PMI and PCI have more general forms that can be used to prove that some property holds for all natural numbers greater than, say, 4. It is a part of the proof to show that the property holds for 5. The next example is such a generalization. It is also a result that is difficult to prove by using the PMI.

Example. Every natural number $n > 1$ has a prime factor.

Proof. Recall that a number p is prime iff its only factors in $\mathbf{N}$ are 1 and p. Let S be $\{n \in \mathbf{N} : n > 1$ and n has a prime factor$\}$. Notice that 1 is not in S, but 2 is in S. Let m be a natural number greater than 1. **Assume** that for all $k \in \{2, \ldots, m - 1\}$, $k \in S$. We must show that $m \in S$. If m has no factors other than 1 and m, then m is prime, and so m has a prime factor *namely, itself* . If m has a factor x other than 1 and m, then $1 < x < m$, so $x \in S$. Therefore, x has a prime factor, which must also be a prime factor of m. In either case, $m \in S$. Therefore,

$S = \{n \in \mathbf{N} : n > 1\}$, and every natural number greater than 1 has a prime factor. ■

Another property characteristic of $\mathbf{N}$ is the well-ordering principle (WOP). The WOP is equivalent to the PMI and the PCI.

Theorem 2.13. The Well-Ordering Principle (WOP). Every nonempty subset of $\mathbf{N}$ has a smallest element.

Proof. Let T be a nonempty subset of $\mathbf{N}$ and suppose T has no smallest element. ⟨*This is a proof by contradiction.*⟩ Let $S = \mathbf{N} - T$.

(i) Since 1 is the smallest element of $\mathbf{N}$ and T has no smallest element, $1 \notin T$. Therefore, 1 must be in S.

(ii) Suppose $n \in S$. Observe that no numbers less than n belong to T, for if some number less than n were in T, then one of those numbers would be the smallest element in T. Now that we know $1, 2, \ldots, n$ are not in T, $n + 1$ cannot be in T, for otherwise it would be the smallest element of T. Therefore, $n + 1 \in S$. By the PMI, $S = \mathbf{N}$. Thus, if T has no smallest elements, we arrive at the contradiction $T = \varnothing$. Therefore, if T is a nonempty subset of $\mathbf{N}$, T has a smallest element. ■

Proofs using the WOP frequently take the form of assuming that some desired property does not hold for all natural numbers. This produces a nonempty set of natural numbers that do not have the property. By working with the **smallest** such number, one can often find a contradiction. As an example of this method, we will prove that for every $n \in \mathbf{N}$, $n + 1 = 1 + n$. Although we would otherwise freely use commutativity of addition in proofs, we will "forget" it for this example.

Example. For every natural number n, $n + 1 = 1 + n$.

Proof. Suppose there exist natural numbers n such that $n + 1 \neq 1 + n$. Let b be the smallest such number. Obviously, $1 + 1 = 1 + 1$, so $b \neq 1$. Thus b must be of the form $b = c + 1$ for some $c \in \mathbf{N}$. ⟨*See Theorem 2.11.*⟩ Then $(c + 1) + 1 \neq 1 + (c + 1)$. By the associative property, $(c + 1) + 1 \neq (1 + c) + 1$. Subtracting 1 ⟨*from the right side*⟩ of each expression, we have $c + 1 \neq 1 + c$. But this is a contradiction, because $c < b$ and b is the smallest such natural number. We conclude that $n + 1 = 1 + n$ for all natural numbers n. ■

As a more interesting application of the WOP, we will prove the division algorithm for natural numbers. An extension of this algorithm to the integers is presented as an exercise (exercise 9).

Theorem 2.14. Let a and $b \in \mathbf{N}$, with $b \leq a$. Then there exist $q \in \mathbf{N}$ and $r \in \mathbf{N} \cup \{0\}$ such that $a = qb + r$, where $0 \leq r < b$. The numbers q and r are the quotient and remainder, respectively.

Proof. Let $T = \{s \in \mathbf{N}: a < sb\}$. By exercise 9, this set is nonempty. Therefore, by the WOP, T contains a smallest element w. Let $q = w - 1$ and $r = a - qb$. Since $a \geq b$, $q \in \mathbf{N}$. Since $q < w$, $a \geq qb$ and $r \geq 0$. By definition of r, $a = qb + r$. Now suppose $r \geq b$. Then $a - (w - 1)b \geq b$, so $a \geq wb$. This contradicts the fact that $a < wb$. Therefore $r < b$. ■

The three principles, the PMI, PCI, and WOP, are valuable tools. They are the first proof methods to turn to whenever it must be shown that a property holds for all natural numbers. In some cases one of the principles is more appropriate than others, but sometimes more than one are suitable.

Exercises 2.4

★ 1. Which of these sets have the inductive property?
   (a) $\varnothing$
   (b) $\{2, 4, 6, 8, 10, \ldots\}$
   (c) $\{1, 2, 4, 5, 6, 7, \ldots\}$
   (d) $\{17\}$

2. Suppose S is inductive. Which of the following must be true?
   (a) If $n + 1 \in S$, then $n \in S$
   ★ (b) If $n \in S$, then $n + 2 \in S$
   (c) If $n + 1 \notin S$, then $n \notin S$
   (d) If $6 \in S$, then $11 \in S$
   ★ (e) $6 \in S$ and $11 \in S$

3. (a) For natural numbers n the value of $n!$ (n factorial) is defined inductively as follows:

$$1! = 1 \quad \text{and} \quad (n + 1)! = (n + 1) \cdot n! \text{ for } n \geq 1.$$

   Find $4!$, $7!$, and $(n + 2)!/n!$.

   ★ (b) For natural numbers n, the nth Fibonacci number is defined inductively by

$$f_1 = 1, f_2 = 1, \quad \text{and} \quad f_{n+2} = f_{n+1} + f_n \text{ for } n \geq 1.$$

   Find f_4, f_7, and $f_{n+3} - f_{n+1}$.

4. Use the PMI to prove the following for all natural numbers n.
   (a) $1 + 4 + 7 + \cdots + (3n - 2) = \frac{1}{2}n(3n - 1)$.
   (b) $3 + 11 + 19 + \cdots + (8n - 5) = 4n^2 - n$.
   (c) $2^1 + 2^2 + 2^3 + \cdots + 2^n = 2^{n+1} - 2$.
   (d) $1 \cdot 1! + 2 \cdot 2! + 3 \cdot 3! + \cdots + n \cdot n! = (n + 1)! - 1$.
   (e) $\dfrac{n^3}{3} + \dfrac{n^5}{5} + \dfrac{7n}{15}$ is an integer.
   (f) $2^n \geq 1 + n$.
   (g) $n^3 + 5n + 6$ is divisible by 3.
   (h) $4^n - 1$ is divisible by 3.
   (i) The sum of the interior angles of a convex polygon of n sides is $(n - 2) \cdot 180°$.
   (j) $3^n > 1 + 2^n$, for all $n > 1$.
   (k) $(n + 1)! > 2^{n+3}$, for $n \geq 5$.
   (l) $4^{n+4} > (n + 4)^4$.
   (m) If a set A has n elements, then $\mathscr{P}(A)$ has 2^n elements.
   (n) $10^n + 3 \cdot 4^{n+2} + 5$ is divisible by 9.
   (o) $f_1 + f_2 + \cdots + f_n = f_{n+2} - 1$ where f_i is the ith Fibonacci number. (See exercise 3(b).)

5. (a) Use the PCI to prove the PMI.
   (b) Use the PCI to prove the WOP.
6. ★ (a) Use the WOP to prove the PMI.
   (b) Use the WOP to prove the PCI.
7. Use the WOP to prove the following:
   (a) If $a > 0$, then for every natural number n, $a^n > 0$.
   (b) For all positive integers a and b, $b \neq a + b$. (*Hint:* Suppose for some a there is b such that $b = a + b$. By the WOP, there is a smallest a_0 such that, for some b, $b = a_0 + b$. Now apply the WOP again.)
8. Use the PCI to prove that the nth Fibonacci number is an integer. (See exercise 3 (b).)
9. Prove that for all a and $b \in \mathbf{N}$, there is a natural number s such that $a < sb$.
10. **The Division Algorithm for Integers:** If a and b are integers, then there exist integers q and r such that $a = bq + r$, where $0 \leq r < |b|$.
    (a) Find q and r when $a = 7$, $b = 2$; $a = -6$, $b = -2$; $a = -5$, $b = 3$; $a = -5$, $b = -3$; $a = 1$, $b = -4$; $a = 0$, $b = 3$.
    (b) Use the WOP to prove the Division Algorithm for Integers.
11. In a certain kind of tournament, every player plays every other player exactly once and either wins or loses. There are no ties. Define a **top** player to be a player who, for every other player x, either beats x or beats a player y who beats x.
    (a) Show that there can be more than one top player.
    ☆ (b) Use the WOP to show that in every such tournament with n players ($n \in \mathbf{N}$), there is at least one top player.
    (c) Use the PMI to show that every n-player tournament has a top player.
12. **Proofs to Grade.**
    ★ (a) **Claim.** For all $n \in \mathbf{N}$, in every set of n horses, all horses have the same color.
    "Proof." Clearly in every set containing exactly 1 horse, all horses have the same color. Now suppose all horses in every set of n horses have the same color. Consider a set of $n + 1$ horses. If we remove one horse, the horses in the remaining set of n horses all have the same color. Now consider a set of n horses obtained by removing some other horse. All horses in this set have the same color. Therefore, all horses in the set of $n + 1$ horses have the same color. By the PMI, the statement is true for every $n \in \mathbf{N}$. ■
    (b) **Claim.** The WOP implies the PCI.
    "Proof." By exercise 5 (a), the PCI implies the PMI. By exercise 6 (a), the WOP implies the PMI. Therefore, the WOP implies the PCI. ■
    (c) **Claim.** For every natural number n, $n^2 + n$ is odd.
    "Proof." The number $n = 1$ is odd. Suppose $n \in \mathbf{N}$ and $n^2 + n$ is odd. Then,

$$(n + 1)^2 + (n + 1) = n^2 + 2n + 1 + n + 1 = (n^2 + n) + (2n + 2)$$

    is the sum of an odd and an even number. Therefore, $(n + 1)^2 + (n + 1)$ is odd. By the PMI, the property that $n^2 + n$ is odd is true for all natural numbers n. ■
    ★ (d) **Claim.** Every natural number n greater than 1 has a prime factor.
    "Proof." Suppose that, whenever $1 < m < n$, m has a prime factor. If n is prime, then n is a prime factor of n. If n is composite, then $n = rs$ where r and s are natural numbers less than n. By the hypothesis of induction, r has a prime factor p. Since p divides r and r divides n, n has a prime factor. In either case, n has a prime factor. By the PCI, every natural number greater than 1 has a prime factor. ■

Relations 3

Given a set of objects, it is often possible to say some of the objects are related. For example, if a and b are integers, we might say that a is related to b if a divides b; or, for the set of all people, two persons are related if they have the same citizenship. In this chapter we will study the idea of "is related to," by making precise the notion of a relation and then concentrating on certain relations called equivalence relations.

SECTION 3.1. CARTESIAN PRODUCTS AND RELATIONS

The study of relations begins with the concept of an **ordered pair,** symbolized as (a, b), which has the property that if either of the coordinates a or b is changed, the ordered pair changes. Thus two ordered pairs (a, b) and (x, y) are *equal* iff $a = x$ and $b = y$.

The adjective "ordered" is used to describe (a, b) because—for example—$(3, 7) \neq (7, 3)$. The ordered pair $(3, 7)$ is not the same as the set $\{3, 7\} = \{7, 3\}$. A more rigorous definition of an ordered pair as a set is given as an exercise.

Generalizing, we can say that the **ordered *n*-tuples** $(a_1, a_2, \ldots, a_n)$ and $(x_1, x_2, \ldots, x_n)$ are equal iff $a_i = x_i$ for $i = 1, 2, \ldots, n$. Thus the 5-tuples $(3, 7, 1, 3, 6)$, $(3, 7, 1, 3, 8)$, and $(7, 3, 1, 3, 6)$ are all different. An ordered 2-tuple is just an ordered pair; an ordered 3-tuple is usually called an ordered triple.

> **Definition.** Let A and B be sets. The set of all ordered pairs having first coordinate in A and second coordinate in B is called the **Cartesian product** (or **cross product**) of A and B and is written $A \times B$. Thus,
>
> $$A \times B = \{(a, b) : a \in A \text{ and } b \in B\}.$$

If $A = \{1, 2\}$ and $B = \{2, 3, 4\}$, then
$A \times B = \{(1, 2), (1, 3), (1, 4), (2, 2), (2, 3), (2, 4)\}$, whereas
$B \times A = \{(2, 1), (2, 2), (3, 1), (3, 2), (4, 1), (4, 2)\}$, which in this example is not equal to $A \times B$.

If $A = \{a, \{x\}, 2\}$ and $B = \{\{a\}, 1\}$, then
$A \times B = \{(a, \{a\}), (a, 1), (\{x\}, \{a\}), (\{x\}, 1), (2, \{a\}), (2, 1)\}$.

The Cartesian product of three or more sets is defined similarly. For example, if A, B, and C are sets, then $A \times B \times C = \{(a, b, c) : a \in A$ and $b \in B$ and $c \in C\}$. The sets $A \times B \times C$ and $A \times (B \times C)$ are different. The first is a set of ordered triples but the second is a set of ordered pairs. In practice this distinction is often not important.

Some useful relationships between the Cartesian product of sets and the other set operations are gathered in the next theorem.

Theorem 3.1. If A, B, C, and D are sets, then

(a) $A \times (B \cup C) = (A \times B) \cup (A \times C)$.
(b) $A \times (B \cap C) = (A \times B) \cap (A \times C)$.
(c) $A \times \varnothing = \varnothing$.
(d) $(A \times B) \cap (C \times D) = (A \cap C) \times (B \cap D)$.
(e) $(A \times B) \cup (C \times D) \subseteq (A \cup C) \times (B \cup D)$.

Proof.

(a) ⟨*Since both $A \times (B \cup C)$ and $(A \times B) \cup (A \times C)$ are sets of ordered pairs, objects of the form (x, y) will be used to show each set is a subset of the other. We use an "iff-argument."*⟩
The ordered pair $(x, y) \in A \times (B \cup C)$
iff $x \in A$ and $y \in B \cup C$
iff $x \in A$ and ($y \in B$ or $y \in C$)
iff ($x \in A$ and $y \in B$) or ($x \in A$ and $y \in C$)
iff $(x, y) \in A \times B$ or $(x, y) \in A \times C$
iff $(x, y) \in (A \times B) \cup (A \times C)$.
Therefore $A \times (B \cup C) = (A \times B) \cup (A \times C)$.

(e) Let $(x, y) \in (A \times B) \cup (C \times D)$. Thus $(x, y) \in A \times B$ or $(x, y) \in C \times D$. If $(x, y) \in A \times B$, then $x \in A$ and $y \in B$. Thus $x \in A \cup C$ and $y \in B \cup D$. ⟨$A \subseteq A \cup C$, $B \subseteq B \cup D$.⟩ Therefore, $(x, y) \in (A \cup C) \times (B \cup D)$. If $(x, y) \in C \times D$, a similar argument shows $(x, y) \in (A \cup C) \times (B \cup D)$. Thus $(A \times B) \cup (C \times D) \subseteq (A \cup C) \times (B \cup D)$.

Parts (b), (c), and (d) are given as exercise 2. ■

Definition. Let A and B be sets. A **relation from A to B** is a subset of $A \times B$. Subsets of $A \times A$ are called **relations on A**. If R is a relation from A to B and sets A and B are understood, we simply say that R is a relation.

If $A = \{1, 2, 3\}$ and $B = \{a, 5, \{b\}, c\}$, then $A \times B$ and $\{(1, a), (2, \{b\}), (2, c)\}$ are relations from A to B. On the other hand, $\{(a, 1), (\{b\}, 2), (c, 2)\}$ and $\{(5, 1), (5, 2)\}$, are relations from B to A.

Since $\varnothing \subseteq A \times B$ for any sets A and B, $\varnothing$ is a relation from A to B. Some care must be taken in mathematical arguments not to overlook this relation.

If R is a relation and if $(a, b) \in R$, we write $a\,R\,b$, and read this as "a is R-related to b." Likewise $a \not R\, b$ means $(a, b) \notin R$.

Let P be the set of all people. Let $L = \{(a, b) \in P \times P$: a has the same last name as $b\}$. Then L is a relation on P. We observe that Sally Brown L Charlie Brown while Buddy Holly $\not L$ Clyde McPhatter.

Let LTE $= \{(x, y) \in \mathbf{R} \times \mathbf{R} : x \leq y\}$. Then $(2, 5) \in$ LTE, so we write 2 LTE 5. Indeed, $(x, y) \in$ LTE iff $x \leq y$, so LTE is the "less than or equal to" relation on $\mathbf{R}$. Thus 2 LTE 5 is consistent with the notation $2 \leq 5$.

Definition. The **domain** of the relation R from A to B is the set

$$\text{Dom}(R) = \{x \in A : \text{there exists } y \in B \text{ such that } x\,R\,y\}.$$

The **range** of the relation R is the set

$$\text{Rng}(R) = \{y \in B : \text{there exists } x \in A \text{ such that } x\,R\,y\}.$$

Thus the domain of R is the set of all first coordinates of ordered pairs in R, and Rng(R) is the set of all second coordinates. By definition, Dom$(R) \subseteq A$ and Rng$(R) \subseteq B$.

For $A = \{s, p, q, \{r\}\}$ and $B = \{a, b, c, d\}$, let R be the relation $\{(p, a), (q, b), (q, c), (\{r\}, b)\}$. Then Dom$(R) = \{p, q, \{r\}\}$ and Rng$(R) = \{a, b, c\}$.

The **graph** of a relation R from A to B is a pictorial representation of the set of ordered pairs in R. When R is an infinite set, it may be that only the significant portion of R can be displayed.

Example. The graph of the relation $R = \{(p, a), (q, b), (q, c), (\{r\}, b)\}$ is shown in figure 3.1.

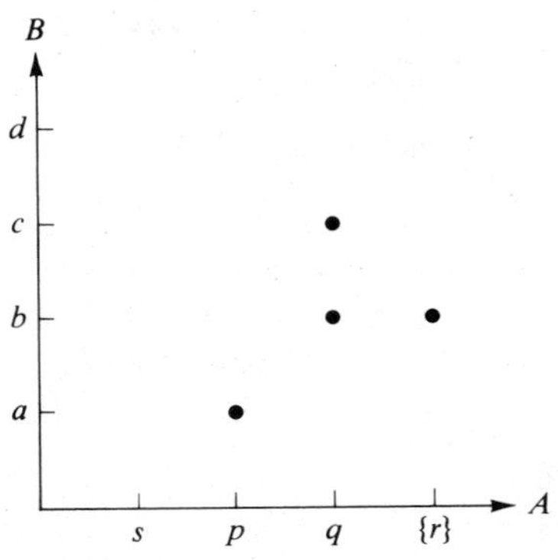

Figure 3.1

Let A be any set. The set $I_A = \{(x, x): x \in A\}$ is called the **identity relation** on A. For $A = \{1, 2\}$, $I_A = \{(1, 1), (2, 2)\}$. Obviously, $\text{Dom}(I_A) = \text{Rng}(I_A) = A$.

Example. For the identity relation on $[-2, \infty)$, we can picture only a portion of the relation. (See figure 3.2.)

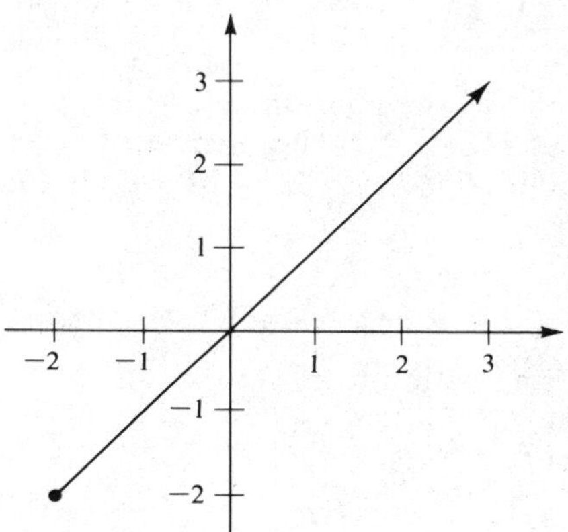

Figure 3.2

Example. The graph of $S = \{(x, y) \in \mathbf{R} \times \mathbf{R}: x^2/324 + y^2/64 \leq 1\}$ is given in figure 3.3.

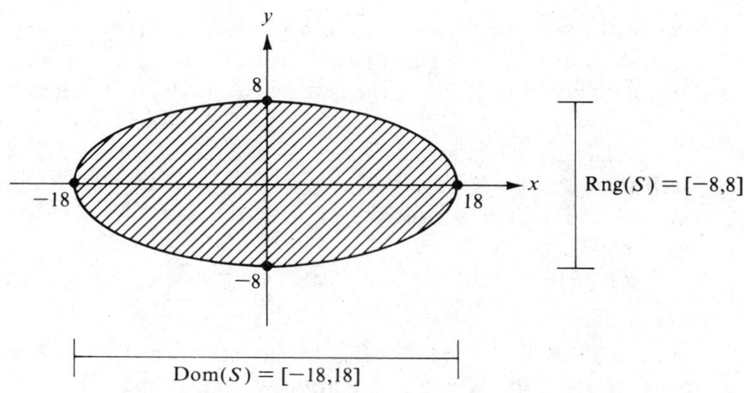

Figure 3.3

Every set of ordered pairs is a relation. If M is any set of ordered pairs, then M is a relation from A to B, where A and B are any sets for which $\text{Dom}(M) \subseteq A$ and $\text{Rng}(M) \subseteq B$.

The two most fundamental methods for constructing new relations from given ones are inversion and composition. Inversion is a matter of switching the order of each pair in a relation.

> **Definition.** If R is a relation from A to B, then the **inverse** of R is
>
> $$R^{-1} = \{(y, x): (x, y) \in R\}.$$

The inverse of the relation $\{(1, b), (1, c), (2, c)\}$ is $\{(b, 1), (c, 1), (c, 2)\}$. For any set A, $I_A^{-1} = I_A$. The inverse of LTE on $\mathbf{R}$ is the relation "greater than or equal to" since x LTE^{-1} y iff y LTE x iff $y \leq x$ iff $x \geq y$.

Theorem 3.2. Let R be a relation from A to B.

(a) R^{-1} is a relation from B to A.
(b) $\text{Dom}(R^{-1}) = \text{Rng}(R)$.
(c) $\text{Rng}(R^{-1}) = \text{Dom}(R)$.

Proof.

(a) Let $(x, y) \in R^{-1}$. ⟨*We show* $(x, y) \in B \times A$.⟩ Then $(y, x) \in R$. Since R is a relation from A to B, $R \subseteq A \times B$. Thus $y \in A$ and $x \in B$. Therefore, $(x, y) \in B \times A$. This proves $R^{-1} \subseteq B \times A$.
(b) $x \in \text{Dom}(R^{-1})$ iff there exists $y \in A$ such that $(x, y) \in R^{-1}$ iff there exists $y \in A$ such that $(y, x) \in R$ iff $x \in \text{Rng}(R)$.
(c) This is similar to part (b). ■

Given a relation from A to B and another from B to C, composition is a method of constructing a relation from A to C. The results developed here are particularly important in the setting of functions in chapter 4.

> **Definition.** Let R be a relation from A to B, and let S be a relation from B to C. The **composite** of R and S is
>
> $$S \circ R = \{(a, c): \text{there exists } b \in B \text{ such that } (a, b) \in R \text{ and } (b, c) \in S\}.$$

Since $S \circ R \subseteq A \times C$, $S \circ R$ is a relation from A to C. It is always true that $\text{Dom}(S \circ R) \subseteq \text{Dom}(R)$, but it is not always true that $\text{Dom}(S \circ R) = \text{Dom}(R)$. We have adopted the right to left notation for $S \circ R$ used in analysis. To determine $S \circ R$, remember that the relation R is from the first set to the second and S is from the second to the third. See figure 3.4.

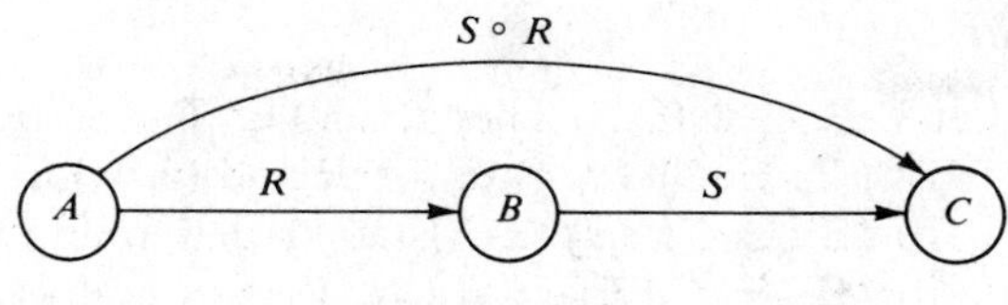

Figure 3.4

Let $A = \{1, 2, 3, 4\}$, $B = \{p, q, r, s\}$, and $C = \{x, y, z\}$. Let $R = \{(1, p), (1, q), (2, q), (3, r), (4, s)\}$ be a relation from A to B, and let $S = \{(p, x), (q, x), (q, y), (s, y)\}$ be a relation from B to C. An element a from A is related to an element c of C under $S \circ R$ if there is at least one "intermediate" element b of B, intermediate in the sense that $(a, b) \in R$ and $(b, c) \in S$. For example, since $(1, p) \in R$ and $(p, x) \in S$, then $(1, x) \in S \circ R$. By determining intermediates we have $S \circ R = \{(1, x), (1, y), (2, x), (2, y), (4, y)\}$.

If R is a relation from A to B, and S is a relation from B to A, then $R \circ S$ and $S \circ R$ are both defined but, owing to the asymmetry in the definition of composition, you should not expect that $R \circ S = S \circ R$. Even when R and S are relations on the same set, it may happen that $R \circ S \neq S \circ R$.

Example. Let $R = \{(x, y) \in \mathbf{R} \times \mathbf{R} : y = x + 1\}$ and let $S = \{(x, y) \in \mathbf{R} \times \mathbf{R} : y = x^2\}$. Then

$$\begin{aligned} R \circ S &= \{(x, y) : (x, z) \in S \text{ and } (z, y) \in R \text{ for some } z \in \mathbf{R}\} \\ &= \{(x, y) : z = x^2 \text{ and } y = z + 1 \text{ for some } z \in \mathbf{R}\} \\ &= \{(x, y) : y = x^2 + 1\}. \end{aligned}$$

$$\begin{aligned} S \circ R &= \{(x, y) : (x, z) \in R \text{ and } (z, y) \in S \text{ for some } z \in \mathbf{R}\} \\ &= \{(x, y) : z = x + 1 \text{ and } y = z^2 \text{ for some } z \in \mathbf{R}\} \\ &= \{(x, y) : y = (x + 1)^2\}. \end{aligned}$$

Clearly, $S \circ R \neq R \circ S$, since $x^2 + 1$ is seldom equal to $(x + 1)^2$.

The last theorem of this section collects several results about inversion, composition, and the identity relation. We prove only part (b) and the first part of (c), leaving the rest as exercise 9.

Theorem 3.3. Suppose A, B, C, and D are sets. Let R be a relation from A to B, S be a relation from B to C, and T be a relation from C to D.

(a) $(R^{-1})^{-1} = R$.
(b) $T \circ (S \circ R) = (T \circ S) \circ R$; that is, composition is associative.
(c) $I_B \circ R = R$ and $R \circ I_A = R$.
(d) $(S \circ R)^{-1} = R^{-1} \circ S^{-1}$.

Proof.

(b) Let $x \in A$, $w \in D$. Then $(x, w) \in T \circ (S \circ R)$
iff $(\exists z \in C)((x, z) \in S \circ R$ and $(z, w) \in T)$
iff $(\exists z \in C)((\exists y \in B)((x, y) \in R$ and $(y, z) \in S)$ and $(z, w) \in T)$
iff $(\exists z \in C)(\exists y \in B)((x, y) \in R$ and $(y, z) \in S$ and $(z, w) \in T)$
iff $(\exists y \in B)(\exists z \in C)((x, y) \in R$ and $(y, z) \in S$ and $(z, w) \in T)$
iff $(\exists y \in B)((x, y) \in R$ and $(\exists z \in C)((y, z) \in S$ and $(z, w) \in T))$
iff $(\exists y \in B)((x, y) \in R$ and $(y, w) \in T \circ S)$
iff $(x, w) \in (T \circ S) \circ R$.
Therefore, $T \circ (S \circ R) = (T \circ S) \circ R$.

(c) ⟨ *We first show that* $I_B \circ R \subseteq R$.⟩ Let $(x, y) \in I_B \circ R$. Then there exists $z \in B$ such that $(x, z) \in R$ and $(z, y) \in I_B$. Since $(z, y) \in I_B$, $z = y$. Thus $(x, y) \in R$ ⟨*since* $(x, y) = (x, z) \in R$⟩. Conversely, let $(p, q) \in R$. Then $(q, q) \in I_B$ and thus $(p, q) \in I_B \circ R$. Thus $I_B \circ R = R$. ■

Exercises 3.1

1. Let A and B be nonempty sets. Prove that $A \times B = B \times A$ iff $A = B$.
2. Complete the proof of Theorem 3.1 by proving
   ★ (b) $A \times (B \cap C) = (A \times B) \cap (A \times C)$.
   (c) $A \times \varnothing = \varnothing$.
   (d) $(A \times B) \cap (C \times D) = (A \cap C) \times (B \cap D)$.
3. Give an example of sets A, B, and C such that $(C \times C) - (A \times B) \neq (C - A) \times (C - B)$.
4. Let T be the relation $\{(3, 1), (2, 3), (3, 5), (2, 2), (1, 6), (2, 6), (1, 2)\}$. Find:
   (a) $\text{Dom}(T)$
   (b) $\text{Rng}(T)$
   (c) T^{-1}
   (d) $(T^{-1})^{-1}$
5. Find the domain and range for the relations:
   ★ (a) $\{(x, y) \in \mathbf{R} \times \mathbf{R} : y = 2x + 1\}$
   (b) $\{(x, y) \in \mathbf{R} \times \mathbf{R} : y = x^2 + 3\}$
   ★ (c) $\{(x, y) \in \mathbf{R} \times \mathbf{R} : y = \sqrt{x - 1}\}$
   (d) $\{(x, y) \in \mathbf{R} \times \mathbf{R} : y = 1/x^2\}$
   ★ (e) $\{(x, y) \in \mathbf{R} \times \mathbf{R} : y < x^2\}$
   (f) $\{(x, y) \in \mathbf{R} \times \mathbf{R} : |x| < 1 \text{ and } y = 3\}$
   (g) $\{(x, y) \in \mathbf{R} \times \mathbf{R} : |x| < 1 \text{ or } y = 3\}$
6. The inverse of $R = \{(x, y) \in \mathbf{R} \times \mathbf{R} : y = 2x + 1\}$ is the relation $R^{-1} = \{(x, y) \in \mathbf{R} \times \mathbf{R} : y = (x - 1)/2\}$. Use this form to give the inverses of the following relations. In (i), (j), and (k), P is the set of all people.
   ★ (a) $R_1 = \{(x, y) \in \mathbf{R} \times \mathbf{R} : y = x\}$
   (b) $R_2 = \{(x, y) \in \mathbf{R} \times \mathbf{R} : y = -5x + 2\}$
   ★ (c) $R_3 = \{(x, y) \in \mathbf{R} \times \mathbf{R} : y = 7x - 10\}$
   (d) $R_4 = \{(x, y) \in \mathbf{R} \times \mathbf{R} : y = x^2 + 2\}$
   ★ (e) $R_5 = \{(x, y) \in \mathbf{R} \times \mathbf{R} : y = -4x^2 + 5\}$
   (f) $R_6 = \{(x, y) \in \mathbf{R} \times \mathbf{R} : y < x + 1\}$
   ★ (g) $R_7 = \{(x, y) \in \mathbf{R} \times \mathbf{R} : y > 3x - 4\}$
   (h) $R_8 = \{(x, y) \in \mathbf{R} \times \mathbf{R} : y = 2x/(x - 2)\}$
   ★ (i) $R_9 = \{(x, y) \in P \times P : y \text{ is the father of } x\}$
   (j) $R_{10} = \{(x, y) \in P \times P : y \text{ is a sibling of } x\}$
   (k) $R_{11} = \{(x, y) \in P \times P : y \text{ loves } x\}$
7. Find the composites for the relations defined in exercise 6.
   ★ (a) $R_1 \circ R_1$
   (b) $R_1 \circ R_2$
   (c) $R_2 \circ R_2$
   ★ (d) $R_2 \circ R_3$
   (e) $R_2 \circ R_4$
   (f) $R_4 \circ R_2$

★ (g) $R_4 \circ R_5$ (h) $R_6 \circ R_2$
(i) $R_6 \circ R_4$ ★ (j) $R_6 \circ R_6$
(k) $R_5 \circ R_5$ (l) $R_8 \circ R_8$
★ (m) $R_3 \circ R_8$ (n) $R_8 \circ R_3$
☆ (o) $R_9 \circ R_9$ (p) $R_{10} \circ R_9$
(q) $R_{11} \circ R_9$

8. Let $A = \{a, b, c, d\}$. Give an example of relations R and S on A such that
(a) $R \circ S \neq S \circ R$ (b) $(S \circ R)^{-1} \neq S^{-1} \circ R^{-1}$
9. Complete the proof of Theorem 3.3.
10. One way to define an ordered pair in terms of sets is to say $(a, b) = \{\{a\}, \{a, b\}\}$. Using this definition, prove that $(a, b) = (x, y)$ iff $a = x$ and $b = y$.
11. Assume ordered pairs have been defined as in exercise 10. Show by example that $(A \times B) \times C = A \times (B \times C)$ is not always true.
12. We may define ordered triples in terms of ordered pairs by saying that $(a, b, c) = ((a, b), c)$. Use this definition to prove that $(a, b, c) = (x, y, z)$ iff $a = x$ and $b = y$ and $c = z$.
13. Generalize and prove a result similar to exercise 12 for n-tuples.
14. **Proofs to Grade.** (See the instructions for exercise 1 on page 23.)

★ (a) **Claim.** $(A \times B) \cup C = (A \times C) \cup (B \times C)$.
"Proof." $x \in (A \times B) \cup C$
iff $x \in A \times B$ or $x \in C$
iff $x \in A$ and $x \in B$ or $x \in C$
iff $x \in A \times C$ or $x \in B \times C$
iff $x \in (A \times C) \cup (B \times C)$ ■

★ (b) **Claim.** If $A \subseteq B$ and $C \subseteq D$, then $A \times C \subseteq B \times D$.
"Proof." Suppose $A \times C \nsubseteq B \times D$. Then there exists a pair $(a, c) \in A \times C$ with $(a, c) \notin B \times D$. But $(a, c) \in A \times C$ implies that $a \in A$ and $c \in C$, whereas $(a, c) \notin B \times D$ implies that $a \notin B$ and $c \notin D$. However, $A \subseteq B$ and $C \subseteq D$, so $a \in B$ and $c \in D$. This is a contradiction. Therefore, $A \times C \subseteq B \times D$. ■

(c) **Claim.** If $A \times B = A \times C$ and $A \neq \varnothing$, then $B = C$.
"Proof." Suppose $A \times B = A \times C$. Then

$$\frac{A \times B}{A} = \frac{A \times C}{A},$$

so $B = C$. ■

★ (d) **Claim.** If $A \times B = A \times C$ and $A \neq \varnothing$, then $B = C$.
"Proof." To show $B = C$, let $b \in B$. Choose any $a \in A$. Then $(a, b) \in A \times B$. But since $A \times B = A \times C$, $(a, b) \in A \times C$. Thus $b \in C$. This proves $B \subseteq C$. A proof of $C \subseteq B$ is similar. Therefore, $B = C$. ■

(e) **Claim.** Let R and S be relations from A to B and from B to C, respectively. Then $S \circ R = (R \circ S)^{-1}$.
"Proof." The ordered pair $(x, y) \in S \circ R$ iff $(y, x) \in R \circ S$ iff $(x, y) \in (R \circ S)^{-1}$. Therefore, $S \circ R = (R \circ S)^{-1}$. ■

(f) **Claim.** Let R be a relation from A to B. Then $I_A \subseteq R^{-1} \circ R$.
"Proof." Let $(x, x) \in I_A$. Choose any $y \in B$ such that $(x, y) \in R$. Then $(y, x) \in R^{-1}$. Thus, $(x, x) \in R^{-1} \circ R$. Therefore, $I_A \subseteq R^{-1} \circ R$. ■

(g) **Claim.** Let R be a relation from A to B. Then $R^{-1} \circ R \subseteq I_A$.
"Proof." Let $(x, y) \in R^{-1} \circ R$. Then for some $z \in B$, $(x, z) \in R$ and $(z, y) \in R^{-1}$. Thus, $(y, z) \in R$. Since $(x, z) \in R$ and $(y, z) \in R$, $x = y$. Thus, $(x, y) = (x, x)$ and $x \in A$, so $(x, y) \in I_A$. ■

SECTION 3.2. EQUIVALENCE RELATIONS

Each of the three properties set forth in the next definitions is important in its own right but relations that possess all three (called equivalence relations) are particularly valuable.

> **Definitions.** Let A be a set and R be a relation on A.
> R is **reflexive** on A iff for all $x \in A$, $x \, R \, x$.
> R is **symmetric** iff for all x and $y \in A$, if $x \, R \, y$, then $y \, R \, x$.
> R is **transitive** iff for all x, y, and $z \in A$, if $x \, R \, y$ and $y \, R \, z$, then $x \, R \, z$.

Only the reflexive property actually asserts that some ordered pairs belong to R. A proof that R is reflexive must then show $x \, R \, x$ for *all* $x \in A$. Since the identity relation on A is the set $I_A = \{(x, x) : x \in A\}$, R is reflexive on A iff $I_A \subseteq R$.

Symmetry and transitivity are defined by conditional sentences and hence most proofs involving these properties are direct proofs. Neither property requires that R contain any ordered pairs. In fact, the empty relation $\varnothing$ is a symmetric and transitive relation on any set A.

For the set $B = \{2, 5, 6, 7\}$, let $S_1 = \{(2, 5), (5, 6), (2, 6)\}$ and $S_2 = \{(2, 5), (2, 2)\}$. Both S_1 and S_2 are transitive relations on B. They are not reflexive on B and not symmetric.

Let R be the relation "is a subset of" on $\mathscr{P}(\mathbf{Z})$, the power set of $\mathbf{Z}$. R is reflexive on $\mathscr{P}(\mathbf{Z})$ since every set is a subset of itself. R is transitive by Theorem 2.3. Notice that $\{1, 2\} \subseteq \{1, 2, 3\}$ but $\{1, 2, 3\} \not\subseteq \{1, 2\}$. Therefore, R is not symmetric.

Let STNR designate the relation $\{(x, y) \in \mathbf{Z} \times \mathbf{Z} : xy > 0\}$ on $\mathbf{Z}$. In this example, x STNR x for all x in $\mathbf{Z}$ except the integer 0; hence the relation STNR is not reflexive on $\mathbf{Z}$. STNR is symmetric since, if x and y are integers and $xy > 0$, then $yx > 0$. STNR is also transitive. To verify this, we assume that x STNR y and y STNR z. Then $xy > 0$ and $yz > 0$. If y is positive, then both x and z are positive; so $xz > 0$. If y is negative, then both x and z are negative; so $xz > 0$. Thus, in either case, x STNR z. Therefore, STNR is symmetric, transitive, and not reflexive on A; whence the acronym that names it.

The identity relation I_A on any set A has all three properties. It is, in fact, the relation "equals," because $x \, I_A \, y$ iff x equals y. Equality is a way of comparing objects according to whether they are the same. Equivalence relations, defined below, are a method of grouping objects according to whether they share a common trait. For example, if T is the set of all triangles, we might say two triangles are alike (equivalent) if they are congruent. This generates the relation $R = \{(x, y) \in T \times T : x \text{ is congruent to } y\}$ on T, which is reflexive on T, symmetric, and transitive. The notion of equivalence, then, is embodied in these three properties.

Definitions. A relation R on a set A is an **equivalence relation on A** iff R is reflexive on A, symmetric, and transitive. For $x \in A$, the **equivalence class of x** determined by R is the set $x/R = \{y \in A : x\,R\,y\}$. This is read "the class of x modulo R" or "x mod R." The set of all equivalence classes is called **A modulo R** and is denoted $A/R = \{x/R : x \in A\}$.

The relation $H = \{(1, 1), (2, 2), (3, 3), (1, 2), (2, 1)\}$ is an equivalence relation on the set $A = \{1, 2, 3\}$. Here $1/H = 2/H = \{1, 2\}$ and $3/H = \{3\}$. Thus $A/H = \{\{1, 2\}, \{3\}\}$.

Let $\Box = \{(x, y) \in \mathbf{R} \times \mathbf{R} : x^2 = y^2\}$. This is the equivalence relation on $\mathbf{R}$ given by $x \Box y$ iff $x^2 = y^2$. In this example, $2/\Box = \{2, -2\}$. Notice that $-\pi/\Box = \{-\pi, \pi\}$, and $0/\Box = \{0\}$. For any $x \in \mathbf{R}$, $x/\Box = \{x, -x\}$.

For the set P of all people, let L be the relation on P given by $x\,L\,y$ iff x and y have the same last name. Then L is an equivalence relation on P. The equivalence class of Charlie Brown modulo L is the set of all people whose last name is Brown. Thus, in addition to Charlie, Charlie Brown/L contains Sally Brown, Buster Brown, Leroy Brown, and all other people who are like Charlie Brown in the sense that they have Brown as a last name. Furthermore, Buster Brown/L = Charlie Brown/L.

Two integers have the same **parity** iff they are either both even or both odd. Let $R = \{(x, y) \in \mathbf{Z} \times \mathbf{Z} : x \text{ and } y \text{ have the same parity}\}$. R is an equivalence relation on $\mathbf{Z}$ with two equivalence classes, the even integers E and the odd integers D. If x is odd, $x/R = D$, while if x is even, $x/R = E$. Thus $\mathbf{Z}/R = \{E, D\}$.

Theorem 3.4. Let R be an equivalence relation on a set A. Then

(a) For all $x \in A$, $x \in x/R$. (Thus $x/R \neq \varnothing$.)
(b) $x/R \subseteq A$ for all $x \in A$.
(c) $A = \bigcup_{x \in A} x/R$.
(d) $x\,R\,y$ iff $x/R = y/R$.
(e) $x \not R\, y$ iff $x/R \cap y/R = \varnothing$.

Proof.

(a) Since R is reflexive on A, $(x, x) \in R$. Thus, $x \in x/R$.
(b) This is from the definition of x/R.
(c) First $\bigcup_{x \in A} x/R \subseteq A$ because each $x/R \subseteq A$. To prove $A \subseteq \bigcup_{x \in A} x/R$, let $t \in A$. By part (a), $t \in t/R \subseteq \bigcup_{x \in A} x/R$. Thus $A = \bigcup_{x \in A} x/R$.
(d) (i) Suppose $x\,R\,y$. To show $x/R = y/R$, we first show $x/R \subseteq y/R$. Let $z \in x/R$. Then $x\,R\,z$. From $x\,R\,y$, by symmetry, $y\,R\,x$. Then, by transitivity, $y\,R\,z$. ⟨*$y\,R\,x$ and $x\,R\,z$ imply $y\,R\,z$.*⟩ Thus $z \in y/R$. The proof that $y/R \subseteq x/R$ is similar.
(ii) Suppose $x/R = y/R$. Since $y \in y/R$, $y \in x/R$. Thus $x\,R\,y$.

(e) (i) If $x/R \cap y/R = \varnothing$, then ⟨*since* $y \in y/R$⟩ $y \notin x/R$. Thus $x \not R y$.
(ii) Finally, we show $x \not R y$ implies $x/R \cap y/R = \varnothing$. ⟨*We prove the contrapositive.*⟩ Suppose $x/R \cap y/R \neq \varnothing$. Let $m \in x/R \cap y/R$. Then $x R m$ and $y R m$. Therefore $x R m$ and $m R y$. Thus $x R y$. ■

The last example we give of an equivalence relation appears in many branches of mathematics. Recall that an integer m divides another integer n iff there exists an integer k such that $n = km$. For a fixed integer $m \neq 0$, let $\equiv_m$ be the relation

$$\{(x, y) \in \mathbf{Z} \times \mathbf{Z} : m \text{ divides } (x - y)\}.$$

We shall see in Theorem 3.5 that $\equiv_m$ is an equivalence relation on $\mathbf{Z}$. The usual notation for $(x, y) \in \equiv_m$ is $\boldsymbol{x \equiv y \ (\text{mod } m)}$, which is read ***x* is congruent to *y* modulo *m*.** We read $x/\equiv_m$ as "the equivalence class of x modulo m" or simply "the class of x mod m." The set of equivalence classes for $\equiv_m$ is denoted $\mathbf{Z}_m$, called "$\mathbf{Z}$ mod m."

As an example, let's concentrate on congruence modulo 3. We see that $4 \equiv 1 \pmod 3$ because 3 divides $4 - 1$. Also, $10 \equiv 16 \pmod 3$, since 3 divides $10 - 16$; that is, $10 - 16 = 3k$ for the integer $k = -2$. However, $5 \not\equiv -6 \pmod 3$ since 3 fails to divide 11.

Next let us calculate all the equivalence classes of $\mathbf{Z}_3$. For $x \in \mathbf{Z}$, the equivalence class of x modulo 3 is $\{y \in \mathbf{Z} : x \equiv y \pmod 3\}$. For 0, we see $0 \equiv 0 \pmod 3$, $3 \equiv 0 \pmod 3$, and $-6 \equiv 0 \pmod 3$. It can be shown that $0/\equiv_3 = \{\ldots, -12, -9, -6, -3, 0, 3, 6, 9, 12, \ldots\}$, the set of all multiples of 3. Similarly, $1/\equiv_3 = \{\ldots, -8, -5, -2, 1, 4, 7, 10, \ldots\}$ and $2/\equiv_3 = \{\ldots, -10, -7, -4, -1, 2, 5, 8, 11, 14, \ldots\}$. There are no other equivalence classes. Thus $\mathbf{Z}_3 = \{0/\equiv_3, 1/\equiv_3, 2/\equiv_3\}$.

Theorem 3.5. The relation $\equiv_m$ is an equivalence relation on the integers. Furthermore, $\mathbf{Z}_m$ has m distinct elements.

Proof. It is clear that $\equiv_m$ is a set of ordered pairs of integers and, hence, is a relation on $\mathbf{Z}$. We shall first show that $\equiv_m$ is an equivalence relation.

(i) To show reflexivity on $\mathbf{Z}$, let x be an integer. We show that $x \equiv x \pmod m$. Since $m \cdot 0 = 0 = x - x$, m divides $x - x$. Thus $\equiv_m$ is reflexive on $\mathbf{Z}$.

(ii) For symmetry, let $x \equiv y \pmod m$. Then m divides $x - y$. Thus there is an integer k, so that $x - y = km$. But this means that $-(x - y) = -(km)$, or that $y - x = (-k)m$. Therefore, m divides $(y - x)$, so that $y \equiv x \pmod m$.

(iii) Suppose $x \equiv y \pmod m$ and $y \equiv z \pmod m$. Thus m divides both $x - y$ and $y - z$. Therefore, there exist integers h and k such that $x - y = hm$ and $y - z = km$. But then $h + k$ is an integer, and $x - z = (x - y) + (y - z) = hm + km = (h + k)m$. Thus m divides $x - z$, so $x \equiv z \pmod m$. Therefore, $\equiv_m$ is transitive.

⟨ *The remainder of the proof shows there are exactly m distinct equivalence classes.* ⟩ We claim

$$\mathbf{Z}_m = \{0/\equiv_m, 1/\equiv_m, 2/\equiv_m, \ldots, (m-1)/\equiv_m\}.$$

First, each $k/\equiv_m$, for $k = 0, 1, 2, \ldots, (m-1)$ is an equivalence class and hence in $\mathbf{Z}_m$. Now suppose $x/\equiv_m \in \mathbf{Z}_m$. By the Division Algorithm, there exist integers q and r such that $x = mq + r$ with $0 \le r \le m - 1$. Thus $mq = x - r$. This implies that $x \equiv r \pmod{m}$ and, therefore, by Theorem 3.4, $x/\equiv_m = r/\equiv_m$. Thus, $x/\equiv_m \in \{0/\equiv_m, 1/\equiv_m, 2/\equiv_m, \ldots, (m-1)/\equiv_m\}$ and the claim is verified.

We will know that $\mathbf{Z}_m$ has exactly m elements when we show that $0/\equiv_m, 1/\equiv_m, 2/\equiv_m, \ldots, (m-1)/\equiv_m$ are all distinct. Suppose $k/\equiv_m = r/\equiv_m$, where $0 \le r \le k \le (m-1)$. Then $k \equiv r \pmod{m}$, and thus m divides $k - r$. But $0 \le k - r \le m - 1$, so $k - r = 0$. Therefore, $k = r$ and the m equivalence classes are distinct. ■

Exercises 3.2

1. Indicate which of the following relations on the given sets are reflexive, which are symmetric, and which are transitive.
   - ★ (a) $\{(1, 2)\}$ on the set $A = \{1, 2\}$.
   - (b) $\le$ on $\mathbf{N}$.
   - (c) $=$ on $\mathbf{N}$.
   - (d) $<$ on $\mathbf{N}$.
   - ★ (e) $\ge$ on $\mathbf{N}$.
   - (f) $\neq$ on $\mathbf{N}$.
   - (g) "divides" on $\mathbf{N}$.
   - (h) $\{(x, y) \in \mathbf{Z} \times \mathbf{Z} : x + y = 10\}$
   - (i) $\perp = \{(\ell, m) : \ell$ and m are lines and ℓ is perpendicular to $m\}$.
   - (j) R, where $(x, y)\, R\, (z, w)$ iff $x + z \le y + w$, on the set $\mathbf{R} \times \mathbf{R}$.
   - ★ (k) S, where $x\, S\, y$ iff x is a sibling of y, on the set P of all people.
   - (l) T, where $(x, y)\, T\, (z, w)$ iff $x + y \le z + w$, on the set $\mathbf{R} \times \mathbf{R}$.
2. Let A be the set $\{1, 2, 3\}$. List the ordered pairs in a relation on A which is
   - ★ (a) not reflexive, not symmetric, and not transitive.
   - (b) reflexive, not symmetric, and not transitive.
   - (c) not reflexive, symmetric, and not transitive.
   - ★ (d) reflexive, symmetric, and not transitive.
   - (e) not reflexive, not symmetric, and transitive.
   - (f) reflexive, not symmetric, and transitive.
   - (g) not reflexive, symmetric, and transitive.
   - (h) reflexive, symmetric, and transitive.
3. ☆ Repeat exercise 2 for relations on $\mathbf{R}$ by sketching graphs of relations with the desired properties.
4. Describe the equivalence classes of each of the following equivalence relations:
   - ★ (a) On $\mathbf{N}$, the relation R given by $a\, R\, b$ iff the prime factorizations of a and b have the same number of 2's.
   - (b) On $\mathbf{R}$, the relation given by $a\, R\, b$ iff $a - b \in \mathbf{Z}$.
   - (c) On the set of all people, the relation "have the same height in centimeters."
   - (d) On $\mathbf{Z}$, the relation R given by $x\, R\, y$ iff $x^2 = y^2$.
5. Calculate the equivalence classes for the relation of
   - ★ (a) congruence modulo 5.
   - (b) congruence modulo 8.
   - (c) congruence modulo 1.
   - (d) congruence modulo 7.

6. The properties of reflexivity, symmetry, and transitivity are related to the identity relation and the operations of inversion and composition. Prove that
   (a) R is reflexive iff $I_A \subseteq R$.
   ★ (b) R is symmetric iff $R = R^{-1}$.
   (c) R is transitive iff $R \circ R \subseteq R$.
7. True or False?
   ★ (a) $4/\equiv_5 = 4/\equiv_6$ (b) $4/\equiv_8 \subseteq 0/\equiv_4$ ★ (c) $2/\equiv_6 \subseteq 2/\equiv_{12}$
   (d) $0/\equiv_{2n} = 0/\equiv_n$ (e) $3/\equiv_5 \cap 4/\equiv_5 = \varnothing$ (f) $8/\equiv_{12} \subseteq 4/\equiv_6$
8. Prove that if R is a symmetric transitive relation with domain A, then R is reflexive on A.
9. ☆ Let L be a relation on a set A that is reflexive and transitive (but not necessarily symmetric). Let R be the relation defined on A by $x\,R\,y$ iff $x\,L\,y$ and $y\,L\,x$. Prove R is an equivalence relation.
10. A relation R on a set A is called
    irreflexive iff for all $x \in A$, $(x, x) \notin R$.
    asymmetric iff for all $x, y \in A$, $(x, y) \in R$ implies $(y, x) \notin R$.
    antisymmetric iff for all $x, y \in A$, $(x, y) \in R$ and $(y, x) \in R$ implies $x = y$.
    (a) Give a mathematical example of a set and a relation which is
        (i) irreflexive and asymmetric
        (ii) antisymmetric
    ★ (b) Prove that if a relation is asymmetric then it is antisymmetric.
11. **Proofs to Grade.**
    (a) **Claim.** If the relation R is symmetric and transitive, it is also reflexive.
    "Proof." Since R is symmetric, if $(x, y) \in R$, then $(y, x) \in R$. Thus $(x, y) \in R$ and $(y, x) \in R$ and since R is transitive, $(x, x) \in R$. Therefore R is reflexive. ■
    (b) **Claim.** If the relations R and S are symmetric, then $R \cap S$ is symmetric.
    "Proof." Suppose $(x, y) \in R \cap S$. Then $(x, y) \in R$ and $(x, y) \in S$. Since R and S are symmetric, $(y, x) \in R$ and $(y, x) \in S$. Therefore, $(y, x) \in R \cap S$. ■
    ★ (c) **Claim.** If the relations R and S are transitive, then $R \cap S$ is transitive.
    "Proof." Suppose $(x, y) \in R \cap S$ and $(y, z) \in R \cap S$. Then $(x, y) \in R$ and $(y, z) \in S$. Therefore, $(x, z) \in R \cap S$. ■

SECTION 3.3. PARTITIONS

If R is an equivalence relation on a set A, Theorem 3.4 tells us that the set A may be thought of as a union of a collection of nonempty subsets that are pairwise disjoint. In other words, imposing an equivalence relation on A results in a set of equivalence classes, so that every element of A is in exactly one class and any two classes are either equal or disjoint. This leads to the concept of a partition of a set.

Definition. Let A be a set. Let $\mathscr{A}$ be a collection of subsets of A. $\mathscr{A}$ is a **partition of A** iff

(i) If $X \in \mathscr{A}$, then $X \neq \varnothing$.
(ii) If $X \in \mathscr{A}$ and $Y \in \mathscr{A}$, then $X = Y$ or $X \cap Y = \varnothing$.
(iii) $\bigcup_{X \in \mathscr{A}} X = A$.

The collections $\{\{0\}, \{1, -1\}, \{2, -2\}, \ldots\}$ and $\{E, D\}$, where E and D are the sets of even and odd integers, respectively, are two different partitions of $\mathbf{Z}$. See figures 3.5 and 3.6.

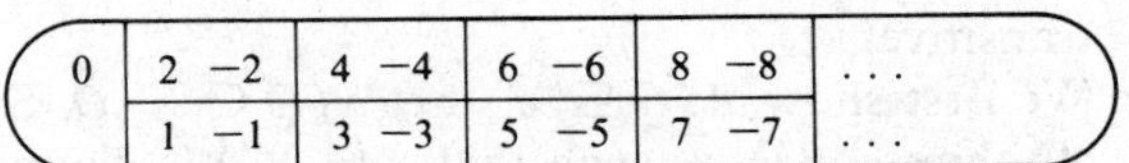

Figure 3.5: A partition of $\mathbf{Z}$

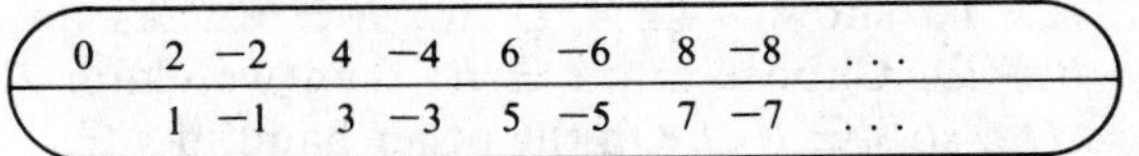

Figure 3.6: A partition of $\mathbf{Z}$

The collection $\{\{1\}, \{2\}, \{3\}, \ldots\}$ is a partition of $\mathbf{N}$. In fact, $\{\{x\} : x \in A\}$ is always a partition of a nonempty set A.

For each $n \in \mathbf{Z}$, let $G_n = [n, n + 1)$. Then $\{G_n : n \in \mathbf{Z}\}$ is a partition of $\mathbf{R}$.

Theorem 3.4 may be restated to say that every equivalence relation on a set determines a partition of that set. We shall soon see that every partition in turn determines an equivalence relation. Thus, each concept may be used to describe the other. This is to our advantage, for we may use either concept, choosing the one that lends itself more readily to the situation at hand.

The method of producing an equivalence relation from a partition is based on the idea that two objects will be said to be equivalent iff they belong to the same member of the partition. For example, let A be the set $\{1, 2, 3, 4, 5\}$ and $\mathscr{B} = \{\{1, 2\}, \{3\}, \{4, 5\}\}$ be a partition of A. To make an equivalence relation Q on A, we note that since $\{1, 2\} \in \mathscr{B}$, 1 is related to 1 and 2, 2 is related to 1 and to 2. Also 3 is related to 3, and so forth. Thus $Q = \{(1, 1), (1, 2), (2, 1), (2, 2), (3, 3), (4, 4), (4, 5), (5, 4), (5, 5)\}$ is the equivalence relation associated with $\mathscr{B}$.

The next theorem asserts that this method of using a partition to define a relation always produces an equivalence relation and, furthermore, that the set of equivalence classes of the relation is the same as the original partition.

Theorem 3.6. Let $\mathscr{B}$ be a partition of the set A. For x and $y \in A$ define $x\,Q\,y$ iff there exists $C \in \mathscr{B}$ such that $x \in C$ and $y \in C$. Then

(a) Q is an equivalence relation on A.
(b) $A/Q = \mathscr{B}$.

Proof.

(a) We prove Q is transitive and leave the proofs of symmetry and reflexivity on A for exercise 3.

Let $x, y, z \in A$. Assume $x\,Q\,y$ and $y\,Q\,z$. Then there are sets C and D in $\mathscr{B}$ such that $x, y \in C$ and $y, z \in D$. Since $\mathscr{B}$ is a partition of A, the sets C and D are either identical or disjoint, but since y is an element of both sets they cannot be disjoint. Hence there is a set C $\langle = D\rangle$ that contains both x and z, so that $x\,Q\,z$. Therefore Q is transitive.

(b) We first show $A/Q \subseteq \mathscr{B}$. Let $x/Q \in A/Q$. Since $\mathscr{B}$ is a partition of A, choose $B \in \mathscr{B}$ such that $x \in B$. We claim $x/Q = B$. If $y \in x/Q$, then $x\,Q\,y$; there is some $C \in \mathscr{B}$ such that $x \in C$ and $y \in C$. Since either $C = B$ or $C \cap B = \varnothing$, and $x \in C \cap B$, $y \in B$. On the other hand, if $y \in B$, then $x\,Q\,y$, and so $y \in x/Q$. Therefore, $x/Q = B$.

To show $\mathscr{B} \subseteq A/Q$, let $B \in \mathscr{B}$. As an element of a partition, $B \neq \varnothing$. Choose any $t \in B$; then we claim $B = t/Q$. If $s \in B$, then $t\,Q\,s$, so $s \in t/Q$. On the other hand, if $s \in t/Q$, then $t\,Q\,s$; so s and t are elements of the same member of $\mathscr{B}$, which must be B. ■

Let $A = \{1, 2, 3, 4\}$ and $\mathscr{B} = \{\{1\}, \{2, 3\}, \{4\}\}$. The equivalence relation Q associated with the partition $\mathscr{B}$ is $\{(1, 1), (2, 2), (2, 3), (3, 2), (3, 3), (4, 4)\}$. The equivalence classes of this relation are $1/Q = \{1\}$, $2/Q = \{2, 3\} = 3/Q$, and $4/Q = \{4\}$, so the set of equivalence classes is precisely $\mathscr{B}$.

For $\mathbf{Z}$, let $\mathscr{A}$ be the partition $\{A_0, A_1, A_2, A_3\}$ where $A_0 = \{\ldots, -8, -4, 0, 4, 8, \ldots\}$, $A_1 = \{\ldots, -7, -3, 1, 5, 9, \ldots\}$, $A_2 = \{\ldots, -6, -2, 2, 6, 10, \ldots\}$, and $A_3 = \{\ldots, -5, -1, 3, 7, 11, \ldots\}$. For integers x and y we see x and y are in the same set A_i if and only if $x = 4k_1 + i$ and $y = 4k_2 + i$, for some integers k_1 and k_2 or, in other words, iff $x - y$ is a multiple of 4. Thus the equivalence relation associated with the partition $\mathscr{A}$ is our old friend, congruence modulo 4.

Exercises 3.3

1. Describe the equivalence relation on $\mathbf{Z}$ determined by the partition $\{E, D\}$ where E is the even integers and D is the odd integers.
2. List the ordered pairs in the equivalence relation on $A = \{1, 2, 3, 4, 5\}$ associated with the partition.
   ★ (a) $\{\{1, 2\}, \{3, 4, 5\}\}$ (b) $\{\{1\}, \{2\}, \{3, 4\}, \{5\}\}$ (c) $\{\{2, 3, 4, 5\}, \{1\}\}$
3. Complete the proof of Theorem 3.6 by proving that if $\mathscr{B}$ is a partition of A, and $x\,Q\,y$ iff there exists $C \in \mathscr{B}$ such that $x \in C$ and $y \in C$, then
   (a) Q is symmetric. (b) Q is reflexive on A.
4. ★ Let R be a relation on a set A which is reflexive and symmetric but not transitive. Let $R(x) = \{y : x\,R\,y\}$. (Note that $R(x)$ is the same as x/R except that R is not an equivalence relation in this exercise.) Does the set $\mathscr{A} = \{R(x) : x \in A\}$ form a partition of A? Prove that your answer is correct.
5. Repeat exercise 4, assuming R is reflexive and transitive but not symmetric.
6. Repeat exercise 4, assuming R is symmetric and transitive but not reflexive.
7. Let A be a set with at least three elements.
   (a) Is there a partition of A with exactly one element?
   ★ (b) If $\mathscr{B} = \{B_1, B_2\}$ is a partition of A with $B_1 \neq B_2$, is $\{\overline{B}_1, \overline{B}_2\}$ a partition of A? Explain. What if $B_1 = B_2$?
   (c) If $\mathscr{B} = \{B_1, B_2, B_3\}$ is a partition of A, is $\{\overline{B}_1, \overline{B}_2, \overline{B}_3\}$ a partition of A? Explain.

(d) If $\mathscr{B} = \{B_1, B_2\}$ is a partition of A, $\mathscr{C}_1$ is a partition of B_1, and $\mathscr{C}_2$ is a partition of B_2, and $B_1 \neq B_2$, prove that $\mathscr{C}_1 \cup \mathscr{C}_2$ is a partition of A.

8. **Proofs to Grade.**

(a) **Claim.** If $\mathscr{A}$ is a partition of a set A and $\mathscr{B}$ is a partition of a set B, then $\mathscr{A} \cup \mathscr{B}$ is a partition of $A \cup B$.

"Proof."

(i) If $X \in \mathscr{A} \cup \mathscr{B}$, then $X \in \mathscr{A}$ or $X \in \mathscr{B}$. In either case $X \neq \varnothing$.

(ii) If $X \in \mathscr{A} \cup \mathscr{B}$ and $Y \in \mathscr{A} \cup \mathscr{B}$, then $X \in \mathscr{A}$ and $Y \in \mathscr{A}$, or $X \in \mathscr{A}$ and $Y \in \mathscr{B}$, or $X \in \mathscr{B}$ and $Y \in \mathscr{A}$, or $X \in \mathscr{B}$ and $Y \in \mathscr{B}$. Since both $\mathscr{A}$ and $\mathscr{B}$ are partitions, in each case either $X = Y$ or $X \cap Y = \varnothing$.

(iii) Since $\bigcup_{X \in \mathscr{A}} X = A$ and $\bigcup_{X \in \mathscr{B}} X = B$, $\bigcup_{X \in \mathscr{A} \cup \mathscr{B}} X = A \cup B$. ■

★ (b) **Claim.** If $\mathscr{B}$ is a partition of A, and if $x\,Q\,y$ iff there exists $C \in \mathscr{B}$ such that $x \in C$ and $y \in C$, then the relation Q is symmetric.

"Proof." First, $x\,Q\,y$ iff there exists $C \in \mathscr{B}$ such that $x \in C$ and $y \in C$. Also, $y\,Q\,x$ iff there exists $C \in \mathscr{B}$ such that $y \in C$ and $x \in C$. Therefore, $x\,Q\,y$ iff $y\,Q\,x$. ■

Functions 4

A function is often thought of as a rule of correspondence between two sets. The statement "Distance is a function of time" means that there is a rule according to which the distance an object has traveled is associated with the time elapsed. The intuitive idea of a function as a rule is useful but not precise enough for a careful study. Unfortunately, the idea of a function as a rule gives the impression that a function must be given by a formula, which is not always the case.

The word "function" was first used by G. W. Leibniz in 1694. J. Bernoulli defined a function as "any expression involving variables and constants" in 1698. The familiar notation $f(x)$ was first used by Euler in 1734. It is only relatively recently that it has become standard practice to treat a function as a relation with special properties. This is possible because the rule that makes an object in one set correspond to an object from a second set may be thought of as producing ordered pairs.

The basic properties of functions, some operations on functions, and induced set functions will be presented in this chapter and used throughout the remainder of the book.

SECTION 4.1. FUNCTIONS AS RELATIONS

Definition. A **function f from A to B** is a relation from A to B that satisfies

(i) $\text{Dom}(f) = A$.
(ii) If $(x, y) \in f$ and $(x, z) \in f$, then $y = z$.

In the case where $A = B$, we say f is a **function on A**.

A function f from A to B is often called a **mapping** of A to B. We write $f: A \to B$, and this is read "f maps A to B" or "f is a function from A to B." As required by the definition, the domain of f is A. The set B is called the

codomain of f. As with any relation, $\text{Rng}(f) = \{v: (\exists u)(u, v) \in f\}$. There is no restriction placed on the sets A and B. They may be sets of numbers, ordered pairs, functions, or even sets of sets of functions.

Let $A = \{1, 2, 3\}$, $B = \{2, 5, 6\}$. The sets

$$\begin{aligned} r_1 &= \{(1, 2), (2, 5), (3, 6), (2, 6)\} \\ r_2 &= \{(1, 2), (2, 6), (3, 5)\} \\ r_3 &= \{(1, 5), (2, 5), (3, 2)\} \\ r_4 &= \{(1, 2), (3, 6)\} \end{aligned}$$

are all relations from A to B. Since $(2, 5)$ and $(2, 6)$ are ordered pairs in r_1 with the same first coordinate, and $5 \neq 6$, r_1 is not a function from A to B. Both r_2 and r_3 satisfy the conditions (i) and (ii), so they are functions from A to B. Since $\text{Dom}(r_4) = \{1, 3\} \neq A$, r_4 is not a function from A to B. However, r_4 is a function from $\{1, 3\}$ to B.

Let $G = \{(x, y) \in \mathbf{N} \times \mathbf{N}: y = x + 2\}$. Then G is a function from $\mathbf{N}$ to $\mathbf{N}$. Also, G is a function from $\mathbf{N}$ to $\mathbf{R}$. Indeed, since $\text{Rng}(G) = \{3, 4, 5, 6, \ldots\}$, G is a function from $\mathbf{N}$ to any set that includes $\{3, 4, 5, 6, \ldots\}$.

Let $H = \{(x, y) \in \mathbf{Z} \times \mathbf{Z}: x^2 + y^2 = 2\}$. Then H is not a function, since $(1, 1) \in H$ and $(1, -1) \in H$, in violation of condition (ii).

In order to verify that a given relation f from A to B is a function from A to B, it must be shown that every element of A appears as a first coordinate of *exactly one* ordered pair in f. The fact that each $a \in A$ is used at least once as a first coordinate makes $\text{Dom}(f) = A$; the fact that a is used *only* once fulfills condition (ii).

It is condition (ii) of the definition that makes f into a rule of correspondence. The rule associated with f is that when $(x, y) \in f$, then y is the unique object that corresponds to x. Having $(1, 6)$ and $(1, 6)$ in a function is allowed (since writing an object twice in a set adds nothing to the set), but having $(1, 6)$ and $(1, 5)$ in f is not allowed, for this would give us two "answers" to the question "What corresponds to 1?"

It is worth noting that the definition of a function says a great deal about first coordinates but almost nothing about the second coordinates of ordered pairs. It may happen that some elements of the codomain are not used as second coordinates, or that some elements of the codomain are used as second coordinates more than once. The relation r_3 above, for example, has $\text{Rng}(r_3) = \{2, 5\} \neq B$, and both $(1, 5)$ and $(2, 5)$ are pairs in r_3. One-to-one and onto functions are functions that satisfy certain conditions on their second coordinates. They will be discussed in a later section.

Definition. Let $f: A \to B$. If $(x, y) \in f$, then we write $f(x) = y$, and say that y is **the value of f at x.** Also, y is the **image** of x under f, and x is a **pre-image** of y under f. Elements of A are sometimes called **arguments** of the function f.

Each argument of a function has exactly one image, but elements of the range may have several pre-images. A function $f\colon A \to B$ may be expressed as $f = \{(x, f(x))\colon x \in A\}$.

It is important to distinguish between the symbols f and $f(x)$. Technically, it is incorrect to speak of "the function $f(x)$." The symbol f denotes a set of ordered pairs, whereas $f(x)$ is simply an element of the range of f. Specifically, $f(x)$ is the element of Rng (f) that corresponds to the element x of Dom(f).

Let $F = \{(x, y) \in \mathbf{Z} \times \mathbf{Z}\colon y = x^2\}$. Then $F\colon \mathbf{Z} \to \mathbf{Z}$. The domain of F is $\mathbf{Z}$, the codomain of F is $\mathbf{Z}$, and $\text{Rng}(F) = \{0, 1, 4, 9, 16, 25, \ldots\}$. The image of 4 is 16, the value of F at -5 is 25, $F(12) = 144$, and both 3 and -3 are pre-images of 9. An element of the codomain that has no pre-image is 6. The function F is given by the rule $F(x) = x^2$.

Let $K\colon \mathbf{R} \to \mathbf{R}$ be defined by $K(x) = 3$ for every $x \in \mathbf{R}$. Then $K = \{(x, 3)\colon x \in \mathbf{R}\}$. This function has range $\{3\}$. The only element of the codomain $\mathbf{R}$ that has a pre-image is 3, and every element of $\mathbf{R}$ is a pre-image of 3. K is an example of a **constant function**—that is, it is a function whose range consists of a single element.

Assume that a universe U has been specified, and that $A \subseteq U$. Define $\chi_A\colon U \to \{0, 1\}$ by

$$\chi_A(x) = \begin{cases} 1 & \text{if } x \in A \\ 0 & \text{if } x \in U - A. \end{cases}$$

Then $\chi_A(x)$ is called the **characteristic function of A.** For example, if $A = [1, 4)$, with the universe being the real numbers, then $\chi_A(x) = 1$ if and only if $1 \leq x < 4$. See figure 4.1.

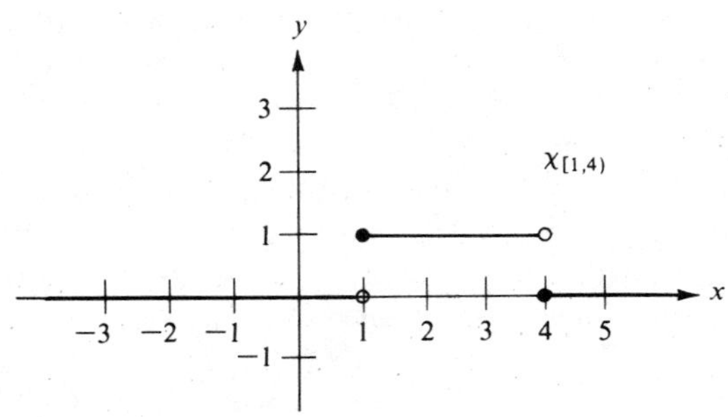

Figure 4.1

If R is an equivalence relation on the set X, then the function from X to X/R that sends each $a \in X$ to a/R is called the **canonical map.** Specifically, if R is the relation of congruence modulo 5 on $\mathbf{Z}$, and f is the canonical map, then the image of 9 is $f(9) = 9/R = \{\ldots, -6, -1, 4, 9, 14, \ldots\}$.

For functions whose domain and range are subsets of $\mathbf{R}$, the domain is often left unspecified. It is assumed then that the domain is the largest possible subset of the reals. For example, since $\sqrt{x+1}$ is a real number iff $x \geq -1$, the domain of the function G given by $G(x) = \sqrt{x+1}$ is $[-1, +\infty)$. The domain of the function H given by $H(x) = 1/(\sin x)$ is $\mathbf{R} - \{k\pi: k \in \mathbf{Z}\}$.

Let A be any set and, for each $x \in A$, let $I_A(x) = x$. Then I_A is a function on A with range A, called the **identity function on A.** In chapter 3 the relation I_A was called the identity relation on A. If $A \subseteq B$, then the function $i: A \to B$ given by $i(x) = x$ for every $x \in A$ is the **inclusion map** from A to B. It is clear that $i = \{(x, x): x \in A\} = I_A$, but i is a function from A to B while I_A is a function from A to A.

Because functions are sets, we already have a definition of function equality: $f = g$ means $f \subseteq g$ and $g \subseteq f$. By this approach, the functions f and g, where $f(x) = (x^2 - 1)/(x + 1)$ and $g(x) = x - 1$, are not equal, because $(-1, -2) \in g$ and $(-1, -2) \notin f$, so $g \nsubseteq f$. A more natural and useful way to express the idea that two functions are equal is to assert that they should have the same domain (so that they act on the same objects) and that for each object in the common domain the functions should agree.

Theorem 4.1. Two functions f and g are equal iff

(i) $\text{Dom}(f) = \text{Dom}(g)$

and

(ii) for all $x \in \text{Dom}(f)$, $f(x) = g(x)$.

Proof. ⟨*We prove that conditions* (i) *and* (ii) *hold when* $f = g$. *The converse is left as an exercise.*⟩ Assume $f = g$.

(i) Let $x \in \text{Dom}(f)$. Then $(x, y) \in f$ for some y and, since $f = g$, $(x, y) \in g$. Therefore, $x \in \text{Dom}(g)$. This shows $\text{Dom}(f) \subseteq \text{Dom}(g)$. Similarly, $\text{Dom}(g) \subseteq \text{Dom}(f)$; so $\text{Dom}(f) = \text{Dom}(g)$.

(ii) Suppose $x \in \text{Dom}(f)$. Then, for some y, $(x, y) \in f$. Since $f = g$, $(x, y) \in g$. Therefore, $f(x) = y = g(x)$. ■

Exercises 4.1

1. Which of the following relations are functions? For those relations that are functions indicate the domain and a possible codomain.

★ (a) $R_1 = \{(0, \triangle), (\triangle, \square), (\square, \cap), (\cap, \cup), (\cup, 0)\}$

(b) $R_2 = \{(1, 2), (1, 3), (1, 4), (1, 5), (1, 6)\}$

(c) $R_3 = \{(1, 2), (2, 1)\}$

(d) $R_4 = \{(x, y) \in \mathbf{R} \times \mathbf{R}: x = \sin y\}$

(e) $R_5 = \{(x, y) \in \mathbf{N} \times \mathbf{N}: x \leq y\}$

(f) $R_6 = \{(x, y) \in \mathbf{Z} \times \mathbf{Z}: y^2 = x\}$

★ (g) $R_7 = \{(x, y) \in \mathbf{R} \times \mathbf{R}: y = x^2 + 2x + 1\}$

(h) $R_8 = \{(x, y) \in \mathbf{N} \times \mathbf{N}: y^2 = x\}$

(i) $R_9 = \{(1, 1), (2, 2), (3, 3), (4, 3), (5, 3)\}$

(j) $R_{10} = \{(\varnothing, \{\varnothing\}), (\{\varnothing\}, \varnothing), (\varnothing, \varnothing), (\{\varnothing\}, \{\varnothing\})\}$

2. Identify the domain, range, and a possible codomain for each of the following functions:
★ (a) $\{(x, y) \in \mathbf{R} \times \mathbf{R} : y = 1/(x - 1)\}$
(b) $\{(x, y) \in \mathbf{R} \times \mathbf{R} : y = x^2 + 5\}$
(c) $\{(x, y) \in \mathbf{N} \times \mathbf{N} : y = x + 5\}$
★ (d) $\{(x, y) \in \mathbf{R} \times \mathbf{R} : y = \tan x\}$
(e) $\{(x, y) \in \mathbf{R} \times \mathbf{R} : y = 1/\cos x\}$
(f) $\{(x, y) \in \mathbf{R} \times \mathbf{R} : y = \chi_{\mathbf{N}}(x)\}$
(g) $\{(x, y) \in \mathbf{N} \times \mathbf{N} : y = 13\}$
(h) $\{(x, y) \in \mathbf{R} \times \mathbf{R} : y = (e^x + e^{-x})/2\}$
(i) $\{(x, y) \in \mathbf{R} \times \mathbf{R} : y = (x^2 - 4)/(x - 2)\}$
(j) $\{(x, y) \in \mathbf{Z} \times \mathbf{Z} : y = (x^2 - 4)/(x - 2)\}$
3. For the real function f given by $f(x) = x^2 - 1$
(a) what is the image of 5 under f?
(b) what is a pre-image of 15?
★ (c) find all pre-images of 24.
(d) what argument of f is associated with the value 20?
(e) what is the value of f at -1?
(f) what is a pre-image of -10?
4. Assuming that the domain of each of the following functions is the largest possible subset of $\mathbf{R}$, find the domain and range of
★ (a) $f(x) = \dfrac{x^2 - 7x + 12}{x - 3}$
(b) $f(x) = 2x + 5$
(c) $f(x) = \dfrac{1}{\sqrt{x + \pi}}$
(d) $f(x) = \sqrt{5 - x}$
(e) $f(x) = \sqrt{5 - x} + \sqrt{x - 3}$
(f) $f(x) = \sqrt{x + 2} + \sqrt{-2 - x}$
5. Explain why the following relations are not functions.
(a) $\{(x, y) \in \mathbf{R} \times \mathbf{R} : x^2 = y^2\}$
(b) $\{(x, y) \in \mathbf{R} \times \mathbf{R} : x^2 + y^2 = 1\}$
(c) $\{(x, y) \in \mathbf{R} \times \mathbf{R} : x = \cos y\}$
6. Is the empty relation a function? Explain.
7. Define $\delta: \mathbf{N} \times \mathbf{N} \to \mathbf{N}$ by $\delta(i, j) = 1$ if $i = j$ and $\delta(i, j) = 0$ if $i \neq j$. Prove that δ is a function. (Frequently we write $\delta(i, j)$ as δ_{ij} and call δ Kronecker's delta.)
8. Let U be the universe and $A \subseteq U$ with $A \neq \varnothing$, $A \neq U$. Let χ_A be the characteristic function of A.
★ (a) What is $\{x \in U: \chi_A(x) = 1\}$?
(b) What is $\{x \in U: \chi_A(x) = 0\}$?
(c) What is $\{x \in U: \chi_A(x) = 2\}$?
9. Explain why the functions $f(x) = (9 - x^2)/(x + 3)$ and $g(x) = 3 - x$ are not equal.
10. Prove the converse of Theorem 4.1. That is, prove that if (i) $\text{Dom}(f) = \text{Dom}(g)$ and (ii) for all $x \in \text{Dom}(f)$, $f(x) = g(x)$, then $f = g$.
11. For the canonical map $f: \mathbf{Z} \to \mathbf{Z}_6$, find
★ (a) $f(3)$
(b) the image of 6
(c) a pre-image of $3/\equiv_6$
(d) all pre-images of $1/\equiv_6$
12. **Proofs to Grade.**
(a) **Claim.** The functions $f(x) = 1 + 1/x$ and $g(x) = (x + 1)/x$ are equal.
"Proof." The domain of each function is assumed to be the largest possible subset of $\mathbf{R}$. Thus $\text{Dom}(f) = \text{Dom}(g) = \mathbf{R} - \{0\}$. For every $x \in \mathbf{R} - \{0\}$ we have $f(x) = 1 + (1/x) = (x/x) + (1/x) = (x + 1)/x = g(x)$. Therefore, by Theorem 4.1, $f = g$. ■
(b) **Claim.** If $h: A \to B$ and $g: C \to D$, then $h \cup g: A \cup C \to B \cup D$.
"Proof." Suppose $(x, y) \in h \cup g$ and $(x, z) \in h \cup g$. Then $(x, y) \in h$ or $(x, y) \in g$,

and $(x, z) \in h$ or $(x, z) \in g$. If $(x, y) \in h$ and $(x, z) \in h$, then $y = z$. Otherwise, $(x, y) \in g$ and $(x, z) \in g$; so again $y = z$. Therefore, $h \cup g$ is a function. Also, $\text{Dom}(h \cup g) = \text{Dom}(h) \cup \text{Dom}(g) = A \cup C$, so $h \cup g: A \cup C \to B \cup D$. ■

SECTION 4.2. CONSTRUCTIONS OF FUNCTIONS

In this section we consider several methods for constructing new functions from given ones. The operations of composition and inversion of relations were discussed in chapter 3. Since every function is a relation, the operations of composition and inversion are performed in the same way as they were for relations. Thus, if $F: A \to B$ and $G: B \to C$, then the inverse of the function F is the relation $F^{-1} = \{(x, y): (y, x) \in F\}$. Similarly, the composite of the functions F and G is the relation $G \circ F = \{(x, z) \in A \times C$: for some $y \in B$ $(x, y) \in F$ and $(y, z) \in G\}$.

For the function $F = \{(x, y): y = 2x + 1\}$ (where x and y are understood to be real numbers) the inverse of F is

$$\begin{aligned} F^{-1} &= \{(x, y): (y, x) \in F\} \\ &= \{(x, y): x = 2y + 1\} \\ &= \{(x, y): y = (x - 1)/2\}, \end{aligned}$$

which is also a function. However, for the function $G = \{(x, y): y = x^2\}$, we have

$$\begin{aligned} G^{-1} &= \{(x, y): (y, x) \in G\} \\ &= \{(x, y): x = y^2\} \\ &= \{(x, y): y = \pm\sqrt{x}\}, \end{aligned}$$

which is not function. Therefore the **inverse of a function is a relation that need not be a function.**

A composite of the two functions F and G, above, is the relation

$$\begin{aligned} G \circ F &= \{(x, z): \text{for some } y \in \mathbf{R}, (x, y) \in F \text{ and } (y, z) \in G\} \\ &= \{(x, z): (\exists y \in \mathbf{R})(y = 2x + 1 \text{ and } z = y^2)\} \\ &= \{(x, z): z = (2x + 1)^2\}. \end{aligned}$$

Before going on with composition we will take advantage of the fact that each element of the domain of a function has a unique image. This will greatly simplify the notation for composition of functions.

For any functions H and K, $(x, z) \in K \circ H$ iff for some y, $(x, y) \in H$ and $(y, z) \in K$. This can be restated as $H(x) = y$ and $K(y) = z$. This means that $(x, z) \in K \circ H$ iff $z = K(H(x))$. The simplification, then, is that $(K \circ H)(x) = K(H(x))$.

For example, the composite $G \circ F$ for the functions given above could be computed as follows: $(G \circ F)(x) = G(F(x)) = G(2x + 1) = (2x + 1)^2$. If $H(x) = \sin x$ and $K(x) = x^2 + 6x$, then

$(H \circ K)(x) = H(K(x)) = H(x^2 + 6x) = \sin(x^2 + 6x)$ and $(K \circ H)(x) = K(H(x)) = K(\sin x) = \sin^2 x + 6 \sin x$. This example shows that $H \circ K$ and $K \circ H$ need not be equal. Thus, composition of functions is not commutative.

Theorem 4.2. If $F: A \to B$ and $G: B \to C$, then $G \circ F: A \to C$. Thus the composite of functions is a function whose domain is the domain of the first function applied.

Proof. *⟨$G \circ F$ is a relation from A to C, and we know that $\text{Dom}(G \circ F) \subseteq A$ and $\text{Rng}(G \circ F) \subseteq C$. To show $G \circ F$ is a function from A to C, we must show (i) $A \subseteq \text{Dom}(G \circ F)$; and (ii) if $(x, y) \in G \circ F$ and $(x, z) \in G \circ F$, then $y = z$.⟩*

(i) Let $x \in A$. Since $A = \text{Dom}(F)$, there is $b \in B$ such that $F(x) = b$. But $B = \text{Dom}(G)$, so there is $c \in C$ such that $G(b) = c$. Then $c = G(b) = G(F(x)) = (G \circ F)(x)$, so $x \in \text{Dom}(G \circ F)$. Therefore, $A \subseteq \text{Dom}(G \circ F)$.

(ii) Assume $(x, y) \in G \circ F$ and $(x, z) \in G \circ F$. Then there is $u \in B$ such that $(x, u) \in F$ and $(u, y) \in G$; and there is $v \in B$ such that $(x, v) \in F$ and $(v, z) \in G$. Since F is a function, $(x, u) \in F$ and $(x, v) \in F$, $u = v$. Then, because G is a function and $(u, y) \in G$ and $(v, z) = (u, z) \in G$, $y = z$. This shows $G \circ F$ is a function. ■

It has already been proved in chapter 3 that composition of relations is associative, and this result applies to functions as well. Similarly, forming the composite of a function with the appropriate identity function yields the function again. These results are restated here especially for functions in order to emphasize their importance and demonstrate the use of functional notation in their proofs.

Theorem 4.3. The composition of functions is associative. That is, if $f: A \to B$, $g: B \to C$, and $h: C \to D$, then $(h \circ g) \circ f = h \circ (g \circ f)$.

Proof. *⟨The idea behind this proof is the characterization of equal functions in Theorem 4.1. The proof also uses the result from Theorem 4.2 that the domain of a composite is the domain of the first function applied.⟩*
The domain of each function is A, by Theorem 4.2. If $x \in A$, then $((h \circ g) \circ f)(x) = (h \circ g)(f(x)) = h(g(f(x))) = h((g \circ f)(x)) = (h \circ (g \circ f))(x)$. ■

Theorem 4.4. Let $f: A \to B$. Then $f \circ I_A = f$ and $I_B \circ f = f$.

Proof. $\text{Dom}(f \circ I_A) = \text{Dom}(I_A) = A = \text{Dom}(f)$. If $x \in A$, then $(f \circ I_A)(x) = f(I_A(x)) = f(x)$. Therefore, $f \circ I_A = f$. The proof that $I_B \circ f = f$ is left as exercise 4. ■

Theorem 4.5. Let $f: A \to B$ with $\text{Rng}(f) = C$. If f^{-1} is a function, then $f^{-1} \circ f = I_A$ and $f \circ f^{-1} = I_C$.

Proof. Suppose $f\colon A \to B$ and f^{-1} is a function. Then $\text{Dom}(f^{-1} \circ f) = \text{Dom}(f)$ ⟨*by Theorem 4.2*⟩. Thus $\text{Dom}(f) = A = \text{Dom}(I_A)$. Let $x \in A$. From the fact that $(x, f(x)) \in f$, we have $(f(x), x) \in f^{-1}$. Therefore, $(f^{-1} \circ f)(x) = f^{-1}(f(x)) = x = I_A(x)$. This proves that $f^{-1} \circ f = I_A$. The proof of the second part of the theorem is left as exercise 5. ■

It is sometimes desirable to create a new function from a given one by removing some of the original ordered pairs. You may recall having seen this done for the sine function with domain **R**. When this function is restricted to the domain $[-\pi/2, \pi/2]$, the result is usually referred to as the Sine function (with a capital S), abbreviated Sin, and is the principal branch of the sine function. Then Sin $(\pi/3)$ = sin $(\pi/3) = \sqrt{3}/2$, but Sin $(2\pi/3)$ is undefined.

Definition. Let $f\colon A \to B$, and let $D \subseteq A$. The **restriction of f to D,** denoted $f|_D$, is

$$\{(x, y)\colon (x, y) \in f \text{ and } x \in D\}.$$

We note that the restriction of any function to D will be a function with domain D. If f and g are mappings and g is a restriction of f, then we say f is an **extension** of g.

In this notation the function Sin is sin $|_{[-\pi/2,\, \pi/2]}$. The graphs of sine and Sine are shown in figure 4.2.

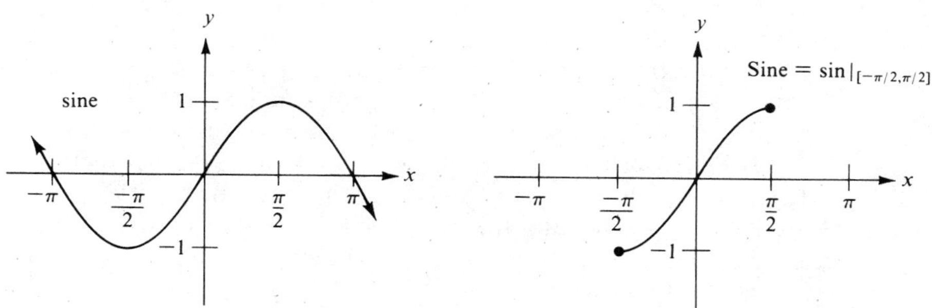

Figure 4.2

Let $A = \{1, 2, 3, 4\}$, $B = \{a, b, c, d\}$, and $g = \{(1, a), (2, a), (3, d), (4, c)\}$. Then $g|_{\{2\}} = \{(2, a)\}$, $g|_A = g$ and $g|_{\{1, 4\}} = \{(1, a), (4, c)\}$.

Let $F\colon \mathbf{R} \to \mathbf{R}$ be given by $F(x) = 2x + 1$. Figure 4.3 shows the graphs of $F|_{[1, 2]}$ and $F|_{\{-2, -1, 0, 1, 2\}}$.

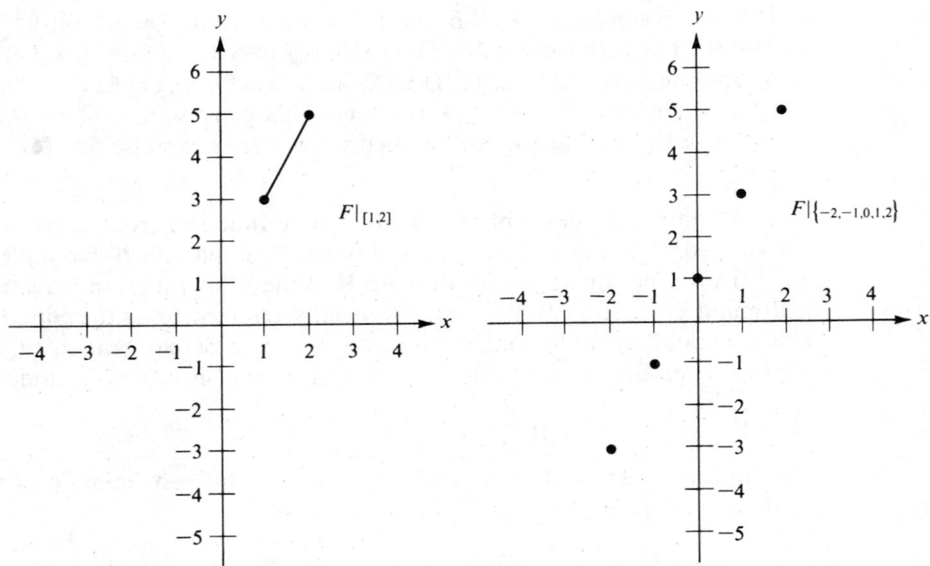

Figure 4.3

Let h and g be functions. Is $h \cap g$ always a function? Is $h \cup g$ always a function? Consider, for example, $h = \{(1, 2), (5, 7), (3, -9)\}$ and $g = \{(1, 8), (5, 7), (4, 8)\}$. Then $h \cap g = \{(5, 7)\}$, which is indeed a function. In general, if x is in the domain of both functions, and $g(x) = h(x) = y$, then $(x, y) \in h \cap g$. An object that is not in both domains or for which $g(x) \neq h(x)$ will not be in $\text{Dom}(h \cap g)$. It turns out that $h \cap g$ is a function (see exercise 8), but this function can just as easily be expressed by restricting the domain of either g or h.

The situation regarding $h \cup g$ is much more interesting and useful. First, for the functions given above, $h \cup g$ is not a function, because (1, 2) and (1, 8) are both in $h \cup g$. If we are careful to be sure that two functions h and g have disjoint domains, however, we can make a new function that is an extension of both h and g by putting them together "piecewise." The proof of Theorem 4.6, which states this result, is left as exercise 9. See exercise 12 for a generalization stating that $h \cup g$ is a function when h and g agree on the intersection of their domains.

Theorem 4.6. Let h and g be functions such that $\text{Dom}(h) = A$, $\text{Dom}(g) = B$, and $A \cap B = \varnothing$. Then $h \cup g$ is a function with domain $A \cup B$.

For example, if $g\colon \{1, 2, 3\} \to \{a, b, c\}$ is the function $\{(1, b), (2, a), (3, c)\}$ and $h = \{(4, d)\}$, then $h \cup g\colon \{1, 2, 3, 4\} \to \{a, b, c, d\}$ is the function $\{(1, b), (2, a), (3, c), (4, d)\}$.

Let h and g be given by $h(x) = x^2$ and $g(x) = 6 - x$. Then $h|_{(-\infty, 2]}$ and $g|_{(2, +\infty)}$ have disjoint domains, so their union f is a function that is an extension of each. It is not an extension of h or g. See figure 4.4.

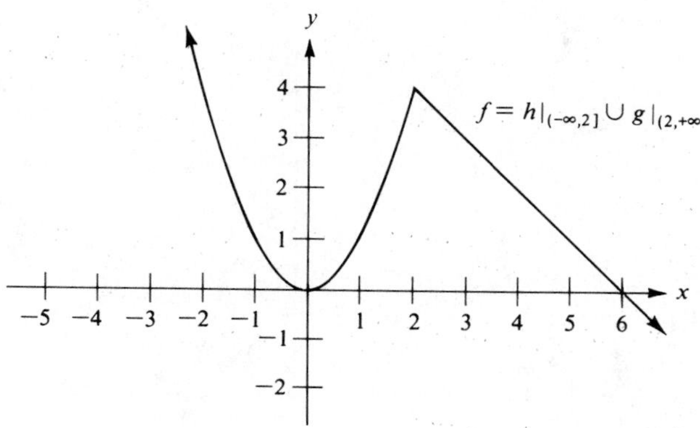

Figure 4.4

The function f can also be described as follows:

$$f(x) = \begin{cases} x^2 & \text{if } x \leq 2 \\ 6 - x & \text{if } x > 2. \end{cases}$$

Functions can also be constructed piecewise from three or more functions. For example, if

$$K(x) = \begin{cases} x + 1 & \text{if } x < -1 \\ \sin \pi x & \text{if } -1 \leq x \leq 0 \\ \dfrac{x + 3}{x - 3} & \text{if } 0 < x < 3 \\ 4 & \text{if } x \geq 4 \end{cases}$$

then K is a function with domain **R**. (See figure 4.5.) To check that the relation given is a function it is only necessary to check that the conditions given on the right are mutually exclusive, so that the corresponding sets are pairwise disjoint. The vertical line test, which says that a graph represents a function as long as no vertical line touches the graph more than once, is useful so long as all the graph can be seen. It is not a rigorous proof that a given relation is a function.

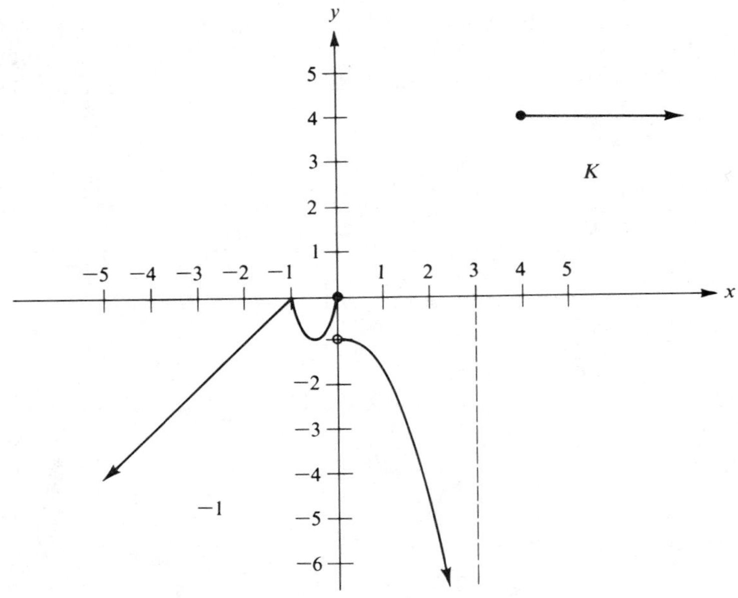

Figure 4.5

Exercises 4.2

1. Find $f \circ g$ and $g \circ f$ for each pair of functions f and g.
   ★ (a) $f(x) = 2x + 5$, $g(x) = 6 - 7x$
   (b) $f(x) = x^2 + 2x$, $g(x) = 2x + 1$
   ★ (c) $f(x) = \sin x$, $g(x) = 2x^2 + 1$
   (d) $f(x) = \tan x$, $g(x) = \sin x$
   (e) $f(x) = \dfrac{x+1}{x+2}$, $g(x) = x^2 + 1$
   (f) $f(x) = 3x + 2$, $g(x) = |x|$

☆ 2. Find the domain and range of each composite in exercise 1.

3. For which of the following functions f is the relation f^{-1} a function? When f^{-1} is a function write an explicit expression for $f^{-1}(x)$.
   ★ (a) $f(x) = 5x + 2$ (b) $f(x) = 2x^2 + 1$ ★ (c) $f(x) = \dfrac{x+1}{x+2}$
   (d) $f(x) = \sin x$ ★ (e) $f(x) = e^{x+3}$ (f) $f(x) = \dfrac{1-x}{-x}$
   (g) $f(x) = \dfrac{1}{1-x}$ (h) $f(x) = -x + 3$ (i) $f(x) = \dfrac{-x}{3x-4}$

4. Prove the remaining part of Theorem 4.4. That is, prove that if $f: A \to B$, then $I_B \circ f = f$.

5. Prove the remaining part of Theorem 4.5. That is, prove that if $f: A \to B$ with $\text{Rng}(f) = C$, and if f^{-1} is a function, then $f \circ f^{-1} = I_C$.

6. Let $f(x) = 4 - 3x$ with domain $\mathbf{R}$ and $A = \{1, 2, 3, 4\}$. Sketch the graphs of the functions $f|_A$, $f|_{[-1,3]}$, $f|_{(2,4]}$, and $f|_{\{6\}}$. What is the range of $f|_{\mathbf{N}}$?

7. Describe two extensions of f with domain $\mathbf{R}$ for the function
   ★ (a) $f = \{(x, y) \in \mathbf{N} \times \mathbf{N} : y = x^2\}$. (b) $f = \{(x, y) \in \mathbf{N} \times \mathbf{N} : y = 3\}$.

8. Prove that, if f and g are functions, then $f \cap g$ is a function by showing that $f \cap g = g|_A$ where $A = \{x: g(x) = f(x)\}$.
9. Prove Theorem 4.6.
10. Let $f(x) = x^2 + 2$ and $g(x) = x + 5$. Describe the function $f|_{(-\infty, 0]} \cup g|_{(0, \infty)}$.
11. For each pair of functions h and g, determine whether $h \cup g$ is a function. In each case sketch a graph of $h \cup g$.

★ (a) $h: (-\infty, 0] \to \mathbf{R}$, $h(x) = 3x + 4$
$g: (0, \infty) \to \mathbf{R}$, $g(x) = \dfrac{1}{x}$

(b) $h: [-1, \infty) \to \mathbf{R}$, $h(x) = x^2 + 1$
$g: (-\infty, -1] \to \mathbf{R}$, $g(x) = x + 3$

(c) $h: (-\infty, 1] \to \mathbf{R}$, $h(x) = |x|$
$g: [0, \infty) \to \mathbf{R}$, $g(x) = 3 - |x - 3|$

(d) $h: (-\infty, 2] \to \mathbf{R}$, $h(x) = \cos x$
$g: [2, \infty) \to \mathbf{R}$, $g(x) = x^2$

(e) $h: (-\infty, 3) \to \mathbf{R}$, $h(x) = 3 - x$
$g: (0, \infty) \to \mathbf{R}$, $g(x) = x + 1$

☆ 12. Let $h: A \to B$, $g: C \to D$ and suppose $E = A \cap C$. Prove $h \cup g$ is a function from $A \cup C$ to $B \cup D$ if and only if $h|_E = g|_E$.

13. **Proofs to Grade.**

★ (a) **Claim.** Let $f: A \to B$. If f^{-1} is a function, then $f^{-1} \circ f = I_A$.
"Proof." Let $(x, y) \in f^{-1} \circ f$. Then there is z such that $(x, z) \in f$ and $(z, y) \in f^{-1}$. But this means that $(z, x) \in f^{-1}$ and $(z, y) \in f^{-1}$. Since f^{-1} is a function, $x = y$. Hence $(x, y) \in f^{-1} \circ f$ implies $(x, y) \in I_A$; that is $f^{-1} \circ f \subseteq I_A$. Now, let $(x, y) \in I_A$. Since $A = \text{Dom}(f)$, there is a $w \in B$ such that $(x, w) \in f$. Hence $(w, x) \in f^{-1}$. But $(x, y) \in I_A$ implies $x = y$ and $(w, y) \in f^{-1}$. But from $(x, w) \in f$ and $(w, y) \in f^{-1}$, we have $(x, y) \in f^{-1} \circ f$. This shows $I_A \subseteq f^{-1} \circ f$. Therefore, $I_A = f^{-1} \circ f$. ■

(b) **Claim.** If f and f^{-1} are functions on A, and $f \circ f = f$, then $f = I_A$.
"Proof." Since $f = f \circ f$, $f^{-1} \circ f = f^{-1} \circ (f \circ f)$. By associativity, we have $f^{-1} \circ f = (f^{-1} \circ f) \circ f$. This gives $I_A = I_A \circ f$. Since $I_A \circ f = f$, we have $I_A = f$. ■

(c) **Claim.** If f, g, and f^{-1} are functions on A, then $g = f^{-1} \circ (g \circ f)$.
"Proof." Using associativity and Theorems 4.4 and 4.5,
$f^{-1} \circ (g \circ f) = f^{-1} \circ (f \circ g) = (f^{-1} \circ f) \circ g = I_A \circ g = g$. ■

SECTION 4.3. ONTO FUNCTIONS; ONE-TO-ONE FUNCTIONS

For every function $f: A \to B$, $\text{Rng}(f) \subseteq B$. It must not be assumed that $\text{Rng}(f) = B$. In the case when the codomain and range are equal, we say the function maps **onto the codomain.**

> **Definition.** A function $f: A \to B$ is **onto *B*** iff $\text{Rng}(f) = B$. We write $f: A \xrightarrow{\text{onto}} B$. A function that maps onto its codomain is also called a **surjection.**

Whether a function f maps onto its codomain or not, it is still correct to say f maps *to* the codomain. It is not good style to say that a function "is onto" without identifying the codomain. There is no harm in simply saying "is onto" when the codomain is clear from the context. As an immediate consequence of the definition, every function maps onto its range.

To prove that a given function $f: A \to B$ is onto a nonempty set B, we choose an arbitrary $y \in B$. We then show $y \in \text{Rng}(f)$ by showing there is $x \in A$ such that $f(x) = y$. This shows $B \subseteq \text{Rng}(f)$, which is sufficient to satisfy the definition, since $\text{Rng}(f) \subseteq B$ is always true.

Example. We will show that $F: \mathbf{R} \to \mathbf{R}$, defined by $F(x) = x + 2$, is onto $\mathbf{R}$ by showing $\mathbf{R} \subseteq \text{Rng}(F)$. To do this, we show that for every $w \in \mathbf{R}$, there is $x \in \mathbf{R}$ such that $F(x) = w$. This x must be chosen such that $F(x) = x + 2 = w$. Then x must be $w - 2$, for with this choice $F(x) = F(w - 2) = (w - 2) + 2 = w$. Hence, F is onto $\mathbf{R}$.

Example. Let $G: \mathbf{R} \to \mathbf{R}$ be defined by $G(x) = x^2 + 1$. Then G is not onto $\mathbf{R}$. To show this, we find an element y in the codomain $\mathbf{R}$ that has no pre-image in the domain $\mathbf{R}$. Let y be -2. Since $x^2 + 1 \geq 1$ for every real number x, there is no $x \in \mathbf{R}$ such that $G(x) = -2$. Hence, G is not onto $\mathbf{R}$. However, G does map $\mathbf{R}$ onto the interval $[1, +\infty)$.

Example. Let $F: \mathbf{N} \times \mathbf{N} \to \mathbf{N}$ be defined by $F(m, n) = 2^{m-1}(2n - 1)$. To show that F is onto $\mathbf{N}$, let $s \in \mathbf{N}$. We must show that there is $(m, n) \in \mathbf{N} \times \mathbf{N}$ such that $F(m, n) = s$. If s is even, then s may be written as $2^j t$, where t is odd. Since t is odd, $t = 2n - 1$ for some $n \in \mathbf{N}$. Choosing $m = j + 1$, we have $F(m, n) = 2^{m-1}(2n - 1) = 2^j t = s$. If s is odd, then $s = 2n - 1$ for some $n \in \mathbf{N}$. For this n and $m = 1$, we find $F(m, n) = 2^0(2n - 1) = s$. Therefore, F is onto $\mathbf{N}$.

The next two theorems relate composition and the property of being onto.

Theorem 4.7. If $f: A \xrightarrow{\text{onto}} B$, and $g: B \xrightarrow{\text{onto}} C$, then $g \circ f: A \xrightarrow{\text{onto}} C$. That is, the composite of onto functions is an onto function.

Proof. Exercise 4. ■

Theorem 4.8. If $f: A \to B$, $g: B \to C$ and $g \circ f: A \xrightarrow{\text{onto}} C$, then g is onto C.

Proof. ⟨*We must show $C \subseteq Rng(g)$.*⟩ Let $c \in C$. Since $g \circ f$ maps onto C, there is $a \in A$ such that $(g \circ f)(a) = c$. Let $b = f(a)$, which is in B. Then $(g \circ f)(a) = g(f(a)) = g(b) = c$. Thus there is $b \in B$ such that $g(b) = c$, and g maps onto C. ■

For a relation f to be a function from A to B, every element of A must appear exactly once as a first coordinate. No restrictions were made for second coordinates, except that they be elements of B. When every element of B appears at least once as a second coordinate, we have called the function onto B. Functions for which every element of B appears at most once as a second coordinate are called one-to-one.

Definition. A function $f: A \to B$ is said to be **one-to-one,** written as $f: A \xrightarrow{1-1} B$, iff $(x, y) \in f$ and $(z, y) \in f$ imply $x = z$. In functional notation, that f is one-to-one means $f(x) = f(z)$ implies $x = z$. A one-to-one function is also called an **injection.**

To prove a given function $f: A \to B$ is one-to-one, we assume x and z are elements of A such that $f(x) = f(z)$. We then show that $x = z$. The alternative, using the contrapositive, is to assume $x \neq z$ and to show that $f(x) \neq f(z)$. To show that f is not one-to-one, it suffices to exhibit two different elements of A with the same image.

Example. The function $F: \mathbf{R} \to \mathbf{R}$ defined by $F(x) = 2x + 1$ is one-to-one. To show this, assume $F(x) = F(z)$. Then $2x + 1 = 2z + 1$. Therefore, $2x = 2z$, so $x = z$.

Example. Let $G(x) = 1/(x^2 + 1)$. We attempt to show G is one-to-one by assuming that $G(x) = G(y)$. Then $1/(x^2 + 1) = 1/(y^2 + 1)$. Therefore, $x^2 + 1 = y^2 + 1$, so $x^2 = y^2$. It does not follow from this that $x = y$. In fact, this failed "proof" suggests a way to find distinct real numbers with equal images. Indeed $G(3) = G(-3) = \frac{1}{10}$. Therefore, G is not one-to-one.

Example. Define $F: \mathbf{N} \times \mathbf{N} \to \mathbf{N}$ by $F(m, n) = 2^{m-1}(2n - 1)$. We will show F is one-to-one. Assume that $F(m, n) = F(r, s)$. We first prove that $m = r$. We may assume that $m \geq r$. ⟨ *If* $m \leq r$ *we could relabel the arguments.* ⟩ From $F(m, n) = F(r, s)$, we have $2^{m-1}(2n - 1) = 2^{r-1}(2s - 1)$, which implies $(2^{m-1}/2^{r-1})(2n - 1) = 2s - 1$. Therefore, $2^{(m-1)-(r-1)}(2n - 1) = 2s - 1$; that is, $2^{m-r}(2n - 1) = 2s - 1$. Since the right side of the equality is odd, the left side is odd. Thus $2^{m-r} = 1$. Therefore, $m - r = 0$, and we conclude that $m = r$.

Dividing both sides of the equation $2^{m-1}(2n - 1) = 2^{r-1}(2s - 1)$ by 2^{m-1} ⟨ $2^{m-1} = 2^{r-1}$ ⟩, we have $2n - 1 = 2s - 1$, which implies $2n = 2s$, or $n = s$. Thus $m = r$ and $n = s$, which gives $(m, n) = (r, s)$. Hence, the function F is one-to-one.

It was observed in the previous section that the inverse of a function is not always a function. The situation is clarified when we understand the connection between inverses and one-to-one functions.

Theorem 4.9. Let $F: A \to B$. F^{-1} is a function from $\text{Rng}(F)$ to A iff F is one-to-one. Furthermore, if F^{-1} is a function, then F^{-1} is one-to-one.

Proof. Assume that $F: A \xrightarrow{1-1} B$. To show that F^{-1} is a function, assume $(x, y) \in F^{-1}$ and $(x, z) \in F^{-1}$. Then $(y, x) \in F$ and $(z, x) \in F$.

Since F is one-to-one, $y = z$. Therefore, if F is one-to-one, F^{-1} is a function. By Theorem 3.2 (b), the domain of F^{-1} is $\text{Rng}(F)$.

Assume now that F^{-1} is a function. To show that F is one-to-one, assume that $(x, y) \in F$ and $(z, y) \in F$. Then $(y, x) \in F^{-1}$ and $(y, z) \in F^{-1}$. Since F^{-1} is a function, $x = z$. Therefore, if F^{-1} is a function, then F is one-to-one.

The proof that if F and F^{-1} are functions, then F^{-1} is one-to-one, is left as exercise 5. ■

We must be careful not to conclude that if $F\colon A \xrightarrow{1-1} B$, then $F^{-1}\colon B \xrightarrow{1-1} A$, since F may not be onto B. Recall that the domain and range of a relation and its inverse are interchanged. Therefore, if $F\colon A \xrightarrow{1-1} B$, then $F^{-1}\colon \text{Rng}(F) \xrightarrow{1-1} A$.

Corollary 4.10. If $F\colon A \xrightarrow[\text{onto}]{1-1} B$, then $F^{-1}\colon B \xrightarrow[\text{onto}]{1-1} A$.

Theorem 4.11. If $f\colon A \xrightarrow{1-1} B$ and $g\colon B \xrightarrow{1-1} C$, then $g \circ f\colon A \xrightarrow{1-1} C$. That is, the composite of one-to-one functions is a one-to-one function.

Proof. Assume that $(g \circ f)(x) = (g \circ f)(z)$; that is, $g(f(x)) = g(f(z))$. Then $f(x) = f(z)$, since g is one-to-one. Then $x = z$, since f is one-to-one. Therefore, $g \circ f$ is one-to-one. ■

There exist functions that are one-to-one but not onto, onto but not one-to-one, neither, and both (see the exercises). A function that is both one-to-one and maps onto its codomain is called a **one-to-one correspondence** or a **bijection.** Combining Theorems 4.7 and 4.11, we have:

Theorem 4.12. If $f\colon A \to B$ and $g\colon B \to C$, and each is a one-to-one correspondence, then $g \circ f\colon A \to C$ is also a one-to-one correspondence.

Analogous to Theorem 4.8 for onto functions, we have the following theorem for one-to-one functions.

Theorem 4.13. If $f\colon A \to B$, $g\colon B \to C$ and if $g \circ f\colon A \xrightarrow{1-1} C$, then $f\colon A \xrightarrow{1-1} B$.

Proof. Exercise 6. ■

We conclude this section with a result that relates the concepts of one-to-one, onto, composition, and inversion, and that also gives a practical method for determining whether two given functions are inverses.

Theorem 4.14. Let $F\colon A \xrightarrow[\text{onto}]{1-1} B$ and $G\colon B \xrightarrow[\text{onto}]{1-1} A$. Then $G = F^{-1}$ iff $G \circ F = I_A$ (or $F \circ G = I_B$).

Proof. If $G = F^{-1}$, then $G \circ F = I_A$ and $F \circ G = I_B$, by Theorem 4.5. Assume now that $G \circ F = I_A$. Then
$G = G \circ I_A = G \circ (F \circ F^{-1}) = (G \circ F) \circ F^{-1} = I_A \circ F^{-1} = F^{-1}$. That $G = F^{-1}$ follows similarly from $F \circ G = I_B$. ■

Let $F(x) = 2x + 1$, and let $G(x) = (x - 1)/2$. Then $F: \mathbf{R} \xrightarrow[\text{onto}]{1-1} \mathbf{R}$ and $G: \mathbf{R} \xrightarrow[\text{onto}]{1-1} \mathbf{R}$. We calculate the two composites
$(G \circ F)(x) = G(F(x)) = G(2x + 1) = [(2x + 1) - 1]/2 = 2x/2 = x$. Also
$(F \circ G)(x) = F(G(x)) = F((x - 1)/2) = 2((x - 1)/2) + 1 = (x - 1) + 1 = x$.
Therefore, $G \circ F = I_{\mathbf{R}}$ and $F \circ G = I_{\mathbf{R}}$. Either computation implies that $G = F^{-1}$.

Constructions of functions by restrictions and unions can also be related to the one-to-one and onto properties of functions. These results will be used in the study of cardinality in chapter 5.

Theorem 4.15.

(a) The restriction of a one-to-one function is one-to-one.
(b) If $h: A \xrightarrow{\text{onto}} C$, $g: B \xrightarrow{\text{onto}} D$, and $A \cap B = \varnothing$, then $h \cup g: A \cup B \xrightarrow{\text{onto}} C \cup D$.
(c) If $h: A \xrightarrow{1-1} C$, $g: B \xrightarrow{1-1} D$, $A \cap B = \varnothing$, and $C \cap D = \varnothing$, then $h \cup g: A \cup B \xrightarrow{1-1} C \cup D$.

Proof. Parts (a) and (b) are left as exercise 7.

(c) Suppose $h: A \xrightarrow{1-1} C$, $g: B \xrightarrow{1-1} D$, $A \cap B = \varnothing$, and $C \cap D = \varnothing$. Then by Theorem 4.6, $h \cup g$ is a function with domain $A \cup B$.
Let $x, y \in A \cup B$. Assume $(h \cup g)(x) = (h \cup g)(y)$.

(i) If $x, y \in A$, then $h(x) = (h \cup g)(x) = (h \cup g)(y) = h(y)$. Since h is one-to-one, $x = y$.
(ii) If $x, y \in B$, then $g(x) = g(y)$, and g is one-to-one; so $x = y$.
(iii) Suppose $x \in A$ and $y \in B$. Then $h(x) = g(y)$ and $h(x) \in C$ and $g(y) \in D$. But $C \cap D = \varnothing$. This case is impossible.
(iv) Similarly, $x \in B$ and $y \in A$ are impossible.

In every possible case, $x = y$. Therefore, $h \cup g$ is one-to-one. ■

Exercises 4.3

1. Which of the following functions map onto their indicated codomains? Prove each of your answers.

★ (a) $f: \mathbf{R} \to \mathbf{R}, f(x) = \frac{1}{2}x + 6$
(b) $f: \mathbf{Z} \to \mathbf{Z}, f(x) = -x + 1000$
★ (c) $f: \mathbf{N} \to \mathbf{N} \times \mathbf{N}, f(x) = (x, x)$
(d) $f: \mathbf{R} \to \mathbf{R}, f(x) = x^3$
(e) $f: \mathbf{R} \to \mathbf{R}, f(x) = \sqrt{x + 5}$
(f) $f: \mathbf{R} \to \mathbf{R}, f(x) = \tan x$
(g) $f: \mathbf{R} \to \mathbf{R}, f(x) = \sin x$
(h) $f: \mathbf{R} \times \mathbf{R} \to \mathbf{R}, f(x, y) = x$
(i) $f: \mathbf{R} \to [-1, 1], f(x) = \cos x$
(j) $f: \mathbf{R} \to [1, \infty), f(x) = x^2 + 1$

☆ 2. Which of the functions in exercise 1 are one-to-one? Prove each of your answers.

3. Let $A = \{1, 2, 3, 4\}$. Describe a codomain B and a function $f: A \to B$ such that f is
★ (a) onto B but not one-to-one. (b) one-to-one but not onto B.
(c) both one-to-one and onto B. (d) neither one-to-one nor onto B.
4. Prove Theorem 4.7.
5. Prove the remaining part of Theorem 4.9. That is, prove that if F and F^{-1} are functions, then F^{-1} is one-to-one.
6. Prove Theorem 4.13.
7. Prove parts (a) and (b) of Theorem 4.15.
8. Give an example of functions $f: A \to B$ and $g: B \to C$ such that
★ (a) f is onto B, but $g \circ f$ is not onto C.
(b) g is onto C, but $g \circ f$ is not onto C.
(c) $g \circ f$ is onto C, but f is not onto B.
(d) f is one-to-one, but $g \circ f$ is not one-to-one.
★ (e) g is one-to-one, but $g \circ f$ is not one-to-one.
(f) $g \circ f$ is one-to-one, but g is not one-to-one.
9. Prove that
(a) $f(x) = \begin{cases} 2 - x & \text{if } x \leq 1 \\ 1/x & \text{if } x > 1 \end{cases}$ is one-to-one but not onto $\mathbf{R}$.
(b) $f(x) = \begin{cases} x + 4 & \text{if } x \leq -2 \\ -x & \text{if } -2 < x < 2 \\ x - 4 & \text{if } x \geq 2 \end{cases}$ is onto $\mathbf{R}$ but not one-to-one.
10. Let A and B be sets and let $S \subseteq A \times B$. Define $f_1: S \to A$ by $f_1(x, y) = x$ for all $(x, y) \in S$. Then f_1 is called the projection of S on A. Similarly, we define $f_2(x, y) = y$ to be the projection of S on B. Give an example to show
★ (a) f_1 need not be one-to-one. (b) f_1 need not be onto A.
(c) f_2 need not be one-to-one. (d) f_2 need not be onto B.
11. Suppose $S: A \to B$ is a function. Then $S \subseteq A \times B$. Let f_1 and f_2 be as in exercise 10.
★ (a) Is $f_1: S \to A$ onto A? (b) Is $f_1: S \to A$ one-to-one?
★ (c) Is $f_2: S \to B$ one-to-one? (d) Is $f_2: S \to B$ onto B?
Prove your answers.
12. **Proofs to Grade.**
(a) **Claim.** If $f: A \xrightarrow{1-1} B$ and $g: B \xrightarrow{1-1} C$, then $g \circ f: A \xrightarrow{1-1} C$ (Theorem 4.11.)
"Proof." We must show that if (x, y) and (z, y) are elements of $g \circ f$, then $x = z$. If $(x, y) \in g \circ f$, then there is $u \in B$ such that $(x, u) \in f$ and $(u, y) \in g$. If $(z, y) \in g \circ f$, then there is $v \in B$ such that $(z, v) \in f$ and $(v, y) \in g$. However, $(u, y) \in g$ and $(v, y) \in g$ imply $u = v$ since g is one-to-one. Then $(x, u) \in f$ and $(z, v) \in f$ and $u = v$; therefore, $x = z$, since f is one-to-one. Hence, (x, y) and (z, y) in $g \circ f$ imply $x = z$. Therefore, $g \circ f$ is one-to-one. ■
(b) **Claim.** The function $f: \mathbf{R} \to \mathbf{R}$ given by $f(x) = 2x + 7$ is one-to-one.
"Proof." Suppose x_1 and x_2 are real numbers with $f(x_1) \neq f(x_2)$. Then $2x_1 + 7 \neq 2x_2 + 7$ and thus $2x_1 \neq 2x_2$. Hence $x_1 \neq x_2$, which shows that f is one-to-one. ■
★ (c) **Claim.** The function $f: \mathbf{R} \to \mathbf{R}$ given by $f(x) = 2x + 7$ is onto $\mathbf{R}$.
"Proof." Suppose f is not onto $\mathbf{R}$. Then there exists $b \in \mathbf{R}$ with $b \notin \text{Rng}(f)$. Thus $b \neq 2x + 7$ for all real numbers x. But $a = \frac{1}{2}(b - 7)$ is a real number and $f(a) = b$. This is a contradiction. Thus f is onto $\mathbf{R}$. ■
★ (d) **Claim.** The function $f: \mathbf{R} \to (-\pi/2, \pi/2)$ given by $f(x) = \arctan(x)$ maps onto $(-\pi/2, \pi/2)$.
"Proof." Let $x \in \mathbf{R}$. Then $f(x) = \arctan(x)$ is a real number such that $-\pi/2 < f(x) < \pi/2$. Thus f maps onto $(-\pi/2, \pi/2)$. ■

SECTION 4.4. INDUCED SET FUNCTIONS

Up to this point, a function f from set A to set B has always been considered "pointwise." That is, we have considered the mapping of individual elements in A to their images in B or else pre-images of elements in B. The next step is to ask about collections of points in A or in B and what corresponds to them in the other set. Every function from A to B *induces* a function that maps subsets of A to subsets of B. Whether or not f^{-1} is a function, there is also a well behaved induced function in the other direction mapping subsets of B to subsets of A.

Definition. Let $f: A \to B$. If $X \subseteq A$, then the **image of X** or **image set of X** is

$$f(X) = \{y \in B: y = f(x) \text{ for some } x \in X\}.$$

If $Y \subseteq B$, then the **inverse image of Y** is

$$f^{-1}(Y) = \{x \in A: f(x) \in Y\}.$$

There need not be any confusion about the fact that the induced functions $f: \mathscr{P}(A) \to \mathscr{P}(B)$ and $f^{-1}: \mathscr{P}(B) \to \mathscr{P}(A)$ have the same names as the function f from A to B and the relation f^{-1} (which may not be a function). The induced functions f and f^{-1} have as domain elements subsets of A and B, respectively. The image set $f(X)$ is just the set of all images of elements of X, and the inverse image $f^{-1}(Y)$ is the set of all pre-images of elements

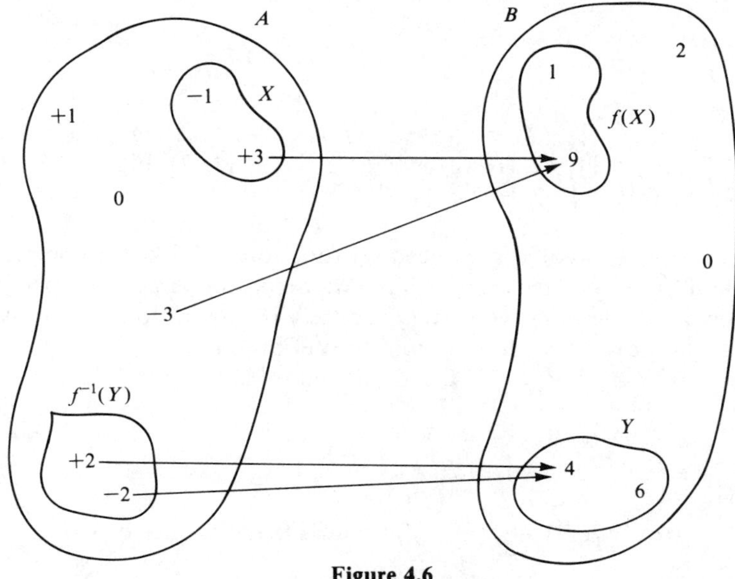

Figure 4.6

of Y. This is illustrated in figure 4.6, where $A = \{0, 1, 2, 3, -1, -2, -3\}$, $B = \{0, 1, 2, 4, 6, 9\}$, and $f: A \to B$ is given by $f(x) = x^2$. The figure shows that $f(\{-1, 3\}) = \{1, 9\}$ and $f^{-1}(\{4, 6\}) = \{2, -2\}$. Also, $f(A) = \{0, 1, 4, 9\}$, $f^{-1}(B) = A$, $f^{-1}(\{6\}) = \varnothing$, and $f(\{3, -3\}) = \{9\}$. Note that f^{-1} is not a function from B to A, so that it would not make sense to consider $f^{-1}(1)$. However, $f^{-1}(\{1\})$ is meaningful and equal to $\{1, -1\}$.

Let $F: \mathbf{R} \to \mathbf{R}$ be given by $F(x) = 2x + 1$. Let $X = \{1, 2, 3\}$. Then $F(X) = \{3, 5, 7\}$ and $F^{-1}(X) = \{0, \frac{1}{2}, 1\}$. Also, $F(\mathbf{N}) = \{3, 5, 7, 9, \ldots\}$ and $F^{-1}(\{2\}) = \{\frac{1}{2}\}$. Figure 4.7 shows that for $D = [1, 2]$, $F(D) = [3, 5]$. Also, $F^{-1}([3, 5]) = [1, 2]$.

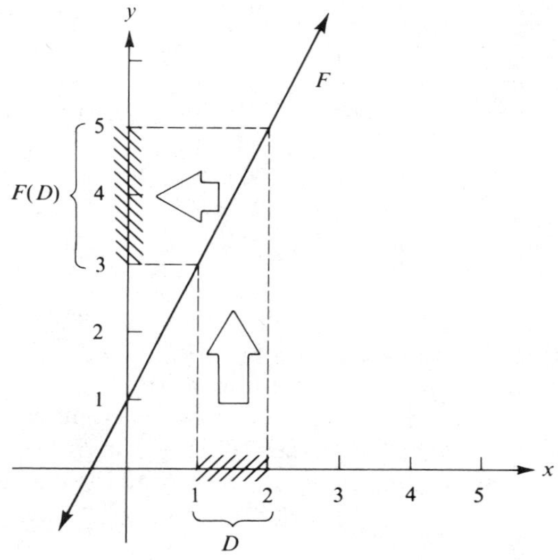

Figure 4.7

Let $f: \mathbf{R} \to \mathbf{R}$ be given by $f(x) = x^2$. Then $f^{-1}([-4, -3]) = \varnothing$ and $f([1, 2]) = [1, 4]$. However, $f^{-1}([1, 4]) \neq [1, 2]$. Figure 4.8 shows that $f^{-1}([1, 4]) = [-2, -1] \cup [1, 2]$.

Proofs involving induced set functions are likely to be more troublesome than others we have seen thus far. Before tackling such proofs it is strongly recommended that you study carefully the definitions to see that each of these facts follows immediately from the definitions.

If $f: A \to B$, $D \subseteq A$, $E \subseteq B$, and $a \in A$, then:

$a \in D \Rightarrow f(a) \in f(D)$
$a \in f^{-1}(E) \Rightarrow f(a) \in E$
$f(a) \in E \Rightarrow a \in f^{-1}(E)$
$f(a) \in f(D) \Rightarrow a \in D$, provided that f is one-to-one.

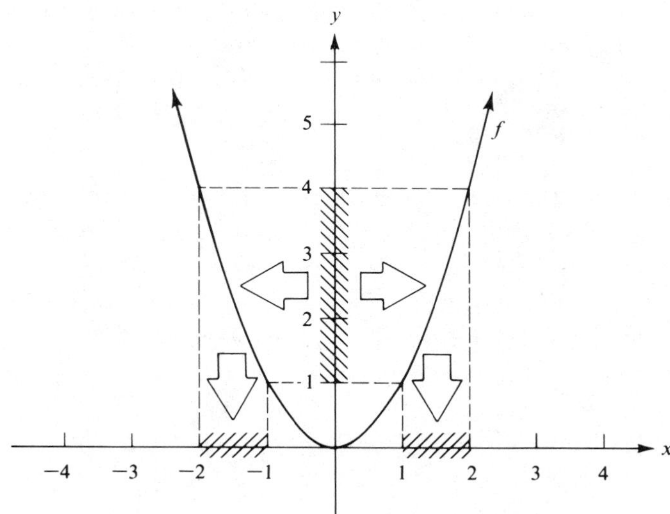

Figure 4.8

You should verify that the last implication is false when f is not one-to-one. Familiarity with these ideas is crucial to understanding induced set functions.

By the algebraic properties of the induced set functions we mean their behavior (properties) as they interact with the operations of union, intersection, and set difference. The next theorems collect results about these properties.

Theorem 4.16. Let $f: A \to B$, and $\{D_\alpha: \alpha \in \Delta\}$ and $\{E_\beta: \beta \in \Gamma\}$ be families of subsets of A and B, respectively. Then

(a) $f(\bigcap_{\alpha \in \Delta} D_\alpha) \subseteq \bigcap_{\alpha \in \Delta} f(D_\alpha)$

(b) $f(\bigcup_{\alpha \in \Delta} D_\alpha) = \bigcup_{\alpha \in \Delta} f(D_\alpha)$

(c) $f^{-1}(\bigcap_{\beta \in \Gamma} E_\beta) = \bigcap_{\beta \in \Gamma} f^{-1}(E_\beta)$

(d) $f^{-1}(\bigcup_{\beta \in \Gamma} E_\beta) = \bigcup_{\beta \in \Gamma} f^{-1}(E_\beta)$.

Proof.

(a) Let $b \in f(\bigcap_{\alpha \in \Delta} D_\alpha)$. Then $b = f(a)$ for some $a \in \bigcap_{\alpha \in \Delta} D_\alpha$. Thus $a \in D_\alpha$ for every $\alpha \in \Delta$. Since $b = f(a)$, $b \in f(D_\alpha)$ for every $\alpha \in \Delta$. Therefore, $b \in \bigcap_{\alpha \in \Delta} f(D_\alpha)$. This proves that $f(\bigcap_{\alpha \in \Delta} D_\alpha) \subseteq \bigcap_{\alpha \in \Delta} f(D_\alpha)$.

Parts (b), (c), and (d) are left as exercise 9. ∎

Theorem 4.17. Let $f: A \to B$ and $E \subseteq B$. Then $f^{-1}(B - E) = A - f^{-1}(E)$.

Proof. Suppose $a \in f^{-1}(B - E)$. Then $f(a) \in B - E$. That is, $f(a) \in B$ and $f(a) \notin E$. Therefore, $a \in A$ and $a \notin f^{-1}(E)$. ⟨ *This is the contrapositive of one of the statements recommended above for study.* ⟩ Thus $a \in A - f^{-1}(E)$. Therefore, $f^{-1}(B - E) \subseteq A - f^{-1}(E)$. The opposite inclusion is left as exercise 10. ■

Theorem 4.18. Let $f: A \to B$, $D \subseteq A$, and $E \subseteq B$. Then

(a) $f(f^{-1}(E)) \subseteq E$.
(b) $E = f(f^{-1}(E))$ iff $E \subseteq \text{Rng}(f)$.
(c) $D \subseteq f^{-1}(f(D))$.
(d) $f^{-1}(f(D)) = D$ iff $f|_{f^{-1}(f(D))}$ is one-to-one.

Proof.

(a) Suppose $b \in f(f^{-1}(E))$. Then there is $a \in f^{-1}(E)$ such that $f(a) = b$. Since $a \in f^{-1}(E)$, $f(a) \in E$. But $f(a) = b$, so $b \in E$. Therefore, $f(f^{-1}(E)) \subseteq E$.
(b) First, suppose $E = f(f^{-1}(E))$. Let $b \in E$. Then $b \in f(f^{-1}(E))$. Thus there is $a \in f^{-1}(E)$ such that $b = f(a)$, so $b \in \text{Rng}(f)$. Therefore, $E \subseteq \text{Rng}(f)$.
Now assume $E \subseteq \text{Rng}(f)$. We know by part (a) that $f(f^{-1}(E)) \subseteq E$, so to prove equality we need only $E \subseteq f(f^{-1}(E))$. Let $b \in E$. Then $b \in \text{Rng}(f)$, so $b = f(a)$ for some $a \in A$. Since $b = f(a) \in E$, $a \in f^{-1}(E)$. Thus $b = f(a)$ and $a \in f^{-1}(E)$, so $b \in f(f^{-1}(E))$. Therefore, $E \subseteq f(f^{-1}(E))$.

Parts (c) and (d) constitute exercise 11. ■

Exercises 4.4

1. Let $A = \{1, 2, 3\}$, $B = \{4, 5, 6\}$, and $h = \{(1, 4), (2, 4), (3, 5)\}$.
★ (a) List the eight ordered pairs in the induced function on $\mathscr{P}(A)$ to $\mathscr{P}(B)$.
(b) List the eight ordered pairs in the induced function on $\mathscr{P}(B)$ to $\mathscr{P}(A)$.
2. Let $f(x) = x^2 + 1$. Find
★ (a) $f([1, 3])$ (b) $f([-1, 0] \cup [2, 4])$ ★ (c) $f^{-1}([-1, 1])$
(d) $f^{-1}([-2, 3])$ (e) $f^{-1}([5, 10])$
3. Let $f(x) = 1 - 2x$. Find
(a) $f(A)$ where $A = \{-1, 0, 1, 2, 3\}$ (b) $f(\mathbf{N})$
(c) $f^{-1}(\mathbf{R})$ (d) $f^{-1}([2, 5])$
(e) $f((1, 4])$
4. Let $f: \mathbf{N} \times \mathbf{N} \to \mathbf{N}$ be given by $f(m, n) = 2^m(2n + 1)$. Find
★ (a) $f^{-1}(A)$ where $A = \{1, 2, 3, 4, 5, 6\}$
(b) $f^{-1}(B)$ where $B = \{4, 6, 8, 10, 12, 14\}$
(c) $f(C)$ where $C = \{(1, 1), (3, 3), (3, 1), (1, 3)\}$
5. Let $f: A \to B$ where $A = \{1, 2, 3, 4, 5, 6\}$, $B = \{p, q, r, s, t, z\}$, and $f = \{(1, p), (2, p), (3, s), (4, t), (5, z), (6, t)\}$. Find
(a) $f^{-1}(\{p, q, s\})$ ★ (b) $f(\{1, 3, 4, 6\})$ (c) $f(\{3, 5\})$
★ (d) $f^{-1}(\{p, r, s, z\})$ (e) $f(f^{-1}(\{p, r, s, z\}))$ (f) $f^{-1}(f(\{1, 4, 5\}))$

6. Let $f: \mathbf{R} - \{0\} \to \mathbf{R}$ be given by $f(x) = x + 1/x$. Find
   (a) $f((0, 2))$ (b) $f((-1, 1])$ ★ (c) $f^{-1}((3, 4])$
   (d) $f^{-1}([0, 1))$ (e) $f(f^{-1}((-4, 10)))$
7. Let $f: \mathbf{R} \to \mathbf{R}$ be given by $f(x) = 10x - x^2$. Find
   ★ (a) $f([1, 6))$ (b) $f^{-1}((0, 21])$ (c) $f^{-1}([52, 54])$
   (d) $f^{-1}([24, 50])$ (e) $f([4, 7])$
8. Let $f: \mathbf{N} \times \mathbf{N} \to \mathbf{N}$ be given by $f(m, n) = 2^m 3^n$. Find
   (a) $f(A \times B)$ where $A = \{1, 2, 3\}$, $B = \{3, 4\}$
   (b) $f^{-1}(\{5, 6, 7, 8, 9, 10\})$
9. Prove parts (b), (c), and (d) of Theorem 4.16.
10. Prove the remaining inclusion of Theorem 4.17. That is, prove that if $f: A \to B$ and $E \subseteq B$, then $A - f^{-1}(E) \subseteq f^{-1}(B - E)$.
11. Prove parts (c) and (d) of Theorem 4.18.
12. Let $f: A \to B$ and let $X, Y \subseteq A$, $U, V \subseteq B$. Prove
    (a) $f(X) \subseteq U$ iff $X \subseteq f^{-1}(U)$. ★ (b) $f(X) - f(Y) \subseteq f(X - Y)$.
    (c) $f^{-1}(U) - f^{-1}(V) = f^{-1}(U - V)$.
13. ☆ Let $f: A \to B$. Prove that, if f is one-to-one, then $f(X) \cap f(Y) = f(X \cap Y)$ for all $X, Y \subseteq A$. Is the converse true? Explain.
14. Let $f: A \to B$. Prove that, if $X \subseteq A$ and f is one-to-one, then $f(A - X) = f(A) - f(X)$.
15. Let $f: A \to B$. Prove that if $X \subseteq A$, $Y \subseteq B$, and f is one-to-one and onto, then $f(X) = Y$ iff $f^{-1}(Y) = X$.
16. Let $f: A \to B$. Consider the function on $\mathcal{P}(A)$ to $\mathcal{P}(B)$ induced by f.
    ★ (a) What condition on f will make the induced function one-to-one?
    (b) What condition on f will make the induced function onto $\mathcal{P}(B)$?
17. Let $f: A \to B$ and $K \subseteq B$. Prove that $f(f^{-1}(K)) = K \cap \operatorname{Rng}(f)$.
18. **Proofs to Grade.**
    ★ (a) **Claim.** If $f: A \to B$ and $X \subseteq A$, then $f^{-1}(f(X)) \subseteq X$.
    "Proof." Let $x \in f^{-1}(f(X))$. Then by definition of f^{-1}, $f(x) \in f(X)$. Therefore, $x \in X$. Thus $f^{-1}(f(X)) \subseteq X$. ■
    (b) **Claim.** If $f: A \to B$ and $X \subseteq A$, then $X \subseteq f^{-1}(f(X))$.
    "Proof." Let $z \in X$. Thus $f(z) \in f(X)$. Therefore, $z \in f^{-1}(f(X))$, which proves the set inclusion. ■
    (c) **Claim.** If $f: A \to B$ and $\{D_\alpha: \alpha \in \Delta\}$ is a family of subsets of A, then $\bigcap_{\alpha \in \Delta} f(D_\alpha) \subseteq f(\bigcap_{\alpha \in \Delta} D_\alpha)$.
    "Proof." Let $y \in \bigcap_{\alpha \in \Delta} f(D_\alpha)$. Then $y \in f(D_\alpha)$ for all α. Thus there exists $x \in D_\alpha$ such that $f(x) = y$, for all α. Then $x \in \bigcap_{\alpha \in \Delta} D_\alpha$ and $f(x) = y$, so $y \in f(\bigcap_{\alpha \in \Delta} D_\alpha)$. Therefore, $\bigcap_{\alpha \in \Delta} f(D_\alpha) \subseteq f(\bigcap_{\alpha \in \Delta} D_\alpha)$. ■

Cardinality 5

How many elements are in the set

$$A = \{\pi, 28, \sqrt{2}, \tfrac{1}{2}, -3, \Delta, \alpha, 0\}?$$

After a short pause you said "eight." Right? Consider for a moment how you arrived at that answer. You probably looked at π and thought "1," then looked at 28 and thought "2," and so on up through 0, which is "8." What you have done is set up a one-to-one correspondence between the set A and the "known" set of eight elements $\{1, 2, 3, 4, 5, 6, 7, 8\}$. Counting the number of elements in sets is essentially a matter of one-to-one correspondences. This process will be extended when we "count" the number of elements in infinite sets in this chapter. Before we begin, here is another counting problem.

A certain shepherd has more than 400 sheep in his flock, but he cannot count beyond 10. Each day he takes his sheep out to graze, and each night he brings them back into the fold. How can he be sure all the sheep have returned? The answer is that he can count them with a one-to-one correspondence. He needs two containers and a pile of pebbles, one pebble for each sheep. When the sheep return in the evening, he transfers pebbles from one container to the other, one at a time for each returning sheep. Whenever there are pebbles left over, he knows there are lost sheep. The solution to the shepherd's problem illustrates the point that even though we have not counted the sheep, we know that the set of missing sheep and the set of leftover pebbles have the same number of elements—because there is a one-to-one correspondence between them.

SECTION 5.1. EQUIVALENT SETS; FINITE SETS

To determine whether two sets have the same number of elements, we see if it is possible to match the elements of the sets in a one-to-one fashion. This idea may be conveniently described in terms of a one-to-one correspondence from one set to the other.

Definition. Two sets A and B are **equivalent** iff there exists a one-to-one function from A onto B. A and B are also said to be **in one-to-one correspondence,** and we write $A \approx B$.

Example. The sets $A = \{5, 8, \phi\}$ and $B = \{r, p, m\}$ are equivalent. The function $f: A \to B$ given by $f(5) = r$, $f(8) = p$, and $f(\phi) = m$ is one of six such functions that verify this.

Example. The set E of even integers is equivalent to D, the set of odd integers. To prove this we employ the function $f: E \to D$ given by $f(x) = x + 1$. The function is one-to-one because $f(x) = f(y)$ implies $x + 1 = y + 1$, which yields $x = y$. Also, f is onto D because if z is any odd integer, then $w = z - 1$ is even and $f(w) = w + 1 = (z - 1) + 1 = z$.

If A and B are not equivalent, we write $A \not\approx B$. For example, $\{p\} \not\approx \{s, t\}$.

Theorem 5.1. The relation $\approx$ is reflexive, symmetric, and transitive. Thus $\approx$ is an equivalence relation on the class of all sets.

Proof. Exercise 1. ■

Theorem 5.2. Suppose A, B, C, and D are sets with $A \approx C$ and $B \approx D$. If A and B are disjoint and C and D are disjoint, then $A \cup B \approx C \cup D$.

Proof. Since $A \approx C$, there is a one-to-one correspondence $h: A \to C$. Similarly, let $g: B \to D$ be a one-to-one correspondence from B to D. Then, by Theorem 4.15, $h \cup g: A \cup B \to C \cup D$ is one-to-one and onto $C \cup D$. Therefore, $A \cup B \approx C \cup D$. ■

We shall use the symbol $\mathbf{N}_k$ to denote the set $\{1, 2, 3, \ldots, k\}$. Each $\mathbf{N}_k$ may be thought of as the standard set with k elements since we will compare the sizes of other sets with them.

Definition. A set S is **finite** iff $S = \varnothing$ or S is equivalent to $\mathbf{N}_k$ for some natural number k. In the case $S = \varnothing$, we say $\varnothing$ **has cardinal number 0** and write $\overline{\overline{\varnothing}} = 0$. If $S \approx \mathbf{N}_k$, then ***S* has cardinal number *k*** and we write $\overline{\overline{S}} = k$. A set S is **infinite** iff it is not finite.

The set $X = \{98.6, c, \pi\}$ is finite and has cardinal number 3. Exhibiting any one-to-one correspondence from X to $\mathbf{N}_3$ will prove this. One such correspondence f is $f(98.6) = 1, f(c) = 2, f(\pi) = 3$.

The set $\{8, 7, 3, 7, 2\}$ is finite and has cardinal number 4, since it is equal to $\{8, 7, 3, 2\}$, which is equivalent to $\mathbf{N}_4$.

The set $\mathbf{N}_k$ is finite and has cardinal number k because the identity function $I_{\mathbf{N}_k}$ is a one-to-one function from $\mathbf{N}_k$ onto $\mathbf{N}_k$.

The cardinality of a finite set is precisely the number of elements in the set. If A is a finite set and B is equivalent to A, then B must also be finite, since, if $A = \varnothing$ or $A \approx \mathbf{N}_k$ for some k, then $B = \varnothing$ or $B \approx \mathbf{N}_k$ by transitivity of $\approx$ (see exercise 2).

The next result of this section will be that every subset of a finite set is finite; our proof uses two lemmas.

Lemma 5.3. If S is a finite set and x is any object, then $S \cup \{x\}$ is finite.

> ***Proof.*** ***Case 1.*** $x \in S$. If $x \in S$, then $S \cup \{x\} = S$, which, by hypothesis, is finite.
> **Case 2.** $x \notin S$. If $S = \varnothing$, then $S \cup \{x\} = \{x\}$, which is equivalent to $\mathbf{N}_1$ and thus finite. If $S \neq \varnothing$, then $S \approx \mathbf{N}_k$ for some k. Also, $\{x\} \approx \{k + 1\}$. Therefore, by Theorem 5.2, $S \cup \{x\} \approx \mathbf{N}_k \cup \{k + 1\} = \mathbf{N}_{k+1}$. This proves $S \cup \{x\}$ is finite. ■

Lemma 5.4. Every subset of $\mathbf{N}_k$ is finite.

> ***Proof.*** Let A be a subset of $\mathbf{N}_k$. ⟨*We prove A is finite by induction on the number k.*⟩
>
> Let $k = 1$. Then either $A = \varnothing$ or $A = \mathbf{N}_1$, both of which are finite.
>
> Assume all subsets of $\mathbf{N}_k$ are finite and let $A \subseteq \mathbf{N}_{k+1}$. Then $A - \{k + 1\}$ is a subset of $\mathbf{N}_k$ and is finite by the hypothesis of induction. If $A = A - \{k + 1\}$, then A is finite. Otherwise $A = (A - \{k + 1\}) \cup \{k + 1\}$ which is finite by Lemma 5.3. ■

Theorem 5.5. Every subset of a finite set is finite.

> ***Proof.*** Assume S is a finite set and $T \subseteq S$. If $T = \varnothing$, then T is finite. Thus we may assume $T \neq \varnothing$ and hence $S \neq \varnothing$. Since $S \approx \mathbf{N}_k$ for some $k \in \mathbf{N}$, let f be a one-to-one correspondence from S onto $\mathbf{N}_k$. Then $f|_T$ is a one-to-one correspondence from T onto the set $f(T)$. (See figure 5.1.) Therefore, $T \approx f(T)$. But $f(T)$ is a subset of $\mathbf{N}_k$ and is finite. Thus T is finite. ■

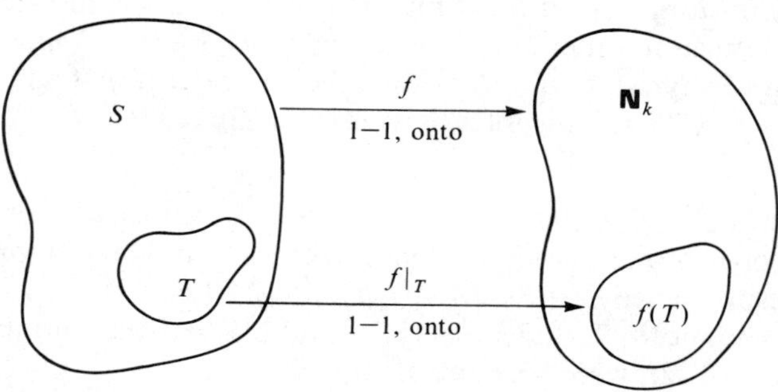

Figure 5.1

The final result of this section is that the union of a finite number of finite sets is finite. To this end the next theorem is a special case: the union of two disjoint finite sets is finite. Its proof is a rigorous development of the idea that, if A has k elements, B has q elements, and $A \cap B = \varnothing$, then $A \cup B$ has $k + q$ elements.

Theorem 5.6. If A and B are finite disjoint sets, then $A \cup B$ is finite and $\overline{\overline{A \cup B}} = \overline{\overline{A}} + \overline{\overline{B}}$.

> ***Proof.*** Suppose A and B are finite sets and $A \cap B = \varnothing$. If $A = \varnothing$, then $A \cup B = B$; if $B = \varnothing$, then $A \cup B = A$. In either case, $A \cup B$ is finite, and since $\overline{\overline{\varnothing}} = 0$, $\overline{\overline{A \cup B}} = \overline{\overline{A}} + \overline{\overline{B}}$. Now, suppose that $A \neq \varnothing$ and $B \neq \varnothing$. Let $A \approx \mathbf{N}_k$ and $B \approx \mathbf{N}_q$, and suppose that $f: A \to \mathbf{N}_k$ and $g: B \to \mathbf{N}_q$ are one-to-one correspondences. Let $H = \{k + 1, k + 2, \ldots, k + q\}$. Then, $h: \mathbf{N}_q \to H$ given by $h(n) = k + n$ is a one-to-one correspondence and, thus, $\mathbf{N}_q \approx H$. Therefore, $B \approx H$ by transitivity. Finally, by Theorem 5.2, $A \cup B \approx \mathbf{N}_k \cup H = \mathbf{N}_{k+q}$, which proves $A \cup B$ is finite and that $\overline{\overline{A \cup B}} = k + q$. ■

Corollary 5.7. (a) If A and B are finite sets, then $A \cup B$ is finite.

(b) If $A_1, A_2, \ldots, A_n$ are finite sets, then $\bigcup_{i=1}^{n} A_i$ is finite.

> ***Proof.*** We prove part (a) and leave part (b) as an exercise in mathematical induction. Assume that both A and B are finite. Since $B - A \subseteq B$, $B - A$ is finite. Thus, by Theorem 5.6, $A \cup B = A \cup (B - A)$ is a finite set. ■

Exercises 5.1

1. Prove Theorem 5.1. That is, show that the relation $\approx$ is reflexive, symmetric, and transitive on the class of all sets.
2. Show that if $A \approx \varnothing$, then $A = \varnothing$.
3. ☆ Show that $A \approx A \times \{x\}$.
4. Which of the following sets are finite?
   - ★ (a) the set of all grains of sand on the earth.
   - (b) the set of all positive integral powers of 2.
   - ★ (c) the set of four-letter words in English.
   - (d) the set of rational numbers.
   - ★ (e) the set of rationals in (0, 1) with denominator 2^k for some $k \in \mathbf{N}$.
   - (f) $\{x \in \mathbf{R} : x^2 + 1 = 0\}$.
   - (g) the set of all turkeys eaten in the year 1620.
   - (h) $\{1, 3, 5\} \times \{2, 4, 6, 8\}$.
   - (i) $\{x \in \mathbf{N} : x \text{ is a prime}\}$.
5. Complete the proof of Corollary 5.7.
6. Let A and B be sets. Prove that
   - ★ (a) if A is finite then $A \cap B$ is finite.
   - (b) if A is infinite, and $A \subseteq B$, then B is infinite.

7. ☆ (a) Prove that for all $k, m \in \mathbf{N}$, $\mathbf{N}_k \times \mathbf{N}_m$ is finite.
   (b) Suppose A and B are finite. Prove $A \times B$ is finite.
8. Define B^A to be the set of all functions from A to B. Show that if A and B are finite, then B^A is finite.
9. If possible, give an example of each:
   (a) an infinite subset of a finite set.
   (b) an infinite collection of finite sets whose union is finite.
   ★ (c) a finite collection of finite sets whose union is infinite.
   (d) finite sets A and B such that $\overline{\overline{A \cup B}} \neq \overline{\overline{A}} + \overline{\overline{B}}$.

☆ 10. Prove that if A and B are finite sets, then $\overline{\overline{A \cup B}} = \overline{\overline{A}} + \overline{\overline{B}} - \overline{\overline{A \cap B}}$.

☆ 11. A "folklore" property of finite sets is the "pigeon hole principle." If a flock of n pigeons comes to roost in a house with r pigeon holes, and $n > r$, at least one hole contains more than one pigeon. Prove the formal version: If $f\colon \mathbf{N}_n \to \mathbf{N}_r$ and $n > r$, then f is not one-to-one.

☆ 12. Use exercise 11 to prove that if A is a proper subset of $\mathbf{N}_k$, then $A \not\approx \mathbf{N}_k$.

13. Use exercise 12 to show a finite set is not equivalent to any of its proper subsets.
14. Prove if A is finite and B is infinite, then $B - A$ is infinite.

☆ 15. Show that if a finite set S has cardinal number n and cardinal number m, then $m = n$.

☆ 16. Prove that if the domain of a function is finite, then the range is finite.

17. **Proofs to Grade.**
   (a) **Claim.** If A and B are finite, then $A \cup B$ is finite.
   "Proof." If A and B are finite, then there exists $m, n \in \mathbf{N}$ such that $A \approx \mathbf{N}_m$, $B \approx \mathbf{N}_n$. Let $f\colon A \xrightarrow[\text{onto}]{1-1} \mathbf{N}_m$ and $h\colon B \xrightarrow[\text{onto}]{1-1} \mathbf{N}_n$. Then $f \cup h\colon A \cup B \xrightarrow[\text{onto}]{1-1} \mathbf{N}_{m+n}$, which shows that $A \cup B \approx \mathbf{N}_{m+n}$. Thus $A \cup B$ is finite. ■
   ★ (b) **Claim.** If S is a finite, nonempty set, then $S \cup \{x\}$ is finite.
   "Proof." Suppose S is finite and nonempty. Then $S \approx \mathbf{N}_k$ for some integer k.
   Case 1. $x \in S$. Then $S \cup \{x\} = S$, so $S \cup \{x\}$ has k elements and is finite.
   Case 2. $x \notin S$. Then $S \cup \{x\} \approx \mathbf{N}_k \cup \{x\} \approx \mathbf{N}_k \cup \mathbf{N}_1 \approx \mathbf{N}_{k+1}$. Thus $S \cup \{x\}$ is finite. ■
   (c) **Claim.** If $A \times B$ is finite, then A is finite.
   "Proof." Choose any $b^* \in B$. Then $A \approx A \times \{b^*\}$. But $A \times \{b^*\} = \{(a, b^*)\colon a \in A\} \subseteq A \times B$. Since $A \times B$ is finite, $A \times \{b^*\}$ is finite. Since A is equivalent to a finite set, A is finite. ■

SECTION 5.2. INFINITE SETS

In the previous section, an infinite set was defined as one that could not be put into a one-to-one correspondence with any $\mathbf{N}_k$. According to the next theorem, the set $\mathbf{N}$ is one such set. Since infinite means not finite, the proof, as might be expected, is by contradiction.

Theorem 5.8. The set of natural numbers is infinite.

Proof. Suppose $\mathbf{N}$ is finite. Clearly $\mathbf{N} \neq \varnothing$. Therefore, for some natural number k, there exists a one-to-one function f from $\mathbf{N}_k$ onto $\mathbf{N}$. ⟨*We will contradict that f is onto* $\mathbf{N}$.⟩ Let $n = f(1) + f(2) + \cdots + f(k) + 1$. Since each $f(i) > 0$, n is a natural number larger than each $f(i)$. Thus $n \neq f(i)$ for any $i \in \mathbf{N}_k$. Hence $n \notin \operatorname{Rng}(f)$. Therefore, f is not onto $\mathbf{N}$, a contradiction. ■

There are many other infinite sets, some, but not all of which, are equivalent to $\mathbf{N}$. It is *not* true that all infinite sets are equivalent.

> **Definition.** A set is **denumerable** iff it is equivalent to $\mathbf{N}$. A denumerable set S has **cardinal number** $\aleph_0$ and we write $\overline{\overline{S}} = \aleph_0$. If a set is finite or denumerable, it is **countable;** otherwise the set is **uncountable.**

The symbol $\aleph$, aleph, is the first symbol in the Hebrew alphabet. The subscript $_0$ (read naught) indicates $\aleph_0$ is the first infinite cardinal number. Other infinite cardinal numbers are associated with uncountable sets.

These definitions along with the ideas of finite and infinite are related in figure 5.2. Note that denumerable sets are those which are countable and infinite.

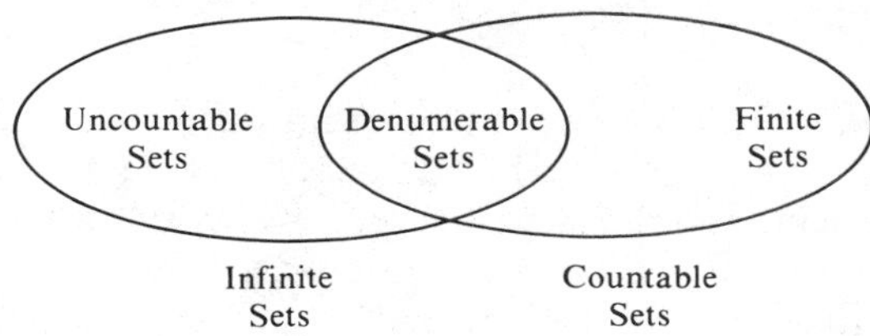

Figure 5.2

The set E^+ of positive even integers is an example of a denumerable set. The function $f\colon \mathbf{N} \to E^+$ defined by setting $f(x) = 2x$ is clearly one-to-one and onto E^+. Of course there are many other one-to-one correspondences; the one given here is the simplest.

What we have shown is that E^+ has the same number of elements as $\mathbf{N}$; that is, E^+ has $\aleph_0$ elements. Thus, although E^+ is a proper subset of $\mathbf{N}$, it could be misleading to say that $\mathbf{N}$ has more elements than E^+. In a later section we will show that every infinite set is equivalent to one of its proper subsets. The next theorem will show that $\mathbf{N}$ and $\mathbf{Z}$ have the same number of elements. This, coupled with our knowledge that $\approx$ is transitive, will show that E^+ and $\mathbf{Z}$ are equivalent. Thus E^+, $\mathbf{N}$, and $\mathbf{Z}$ all have cardinal number $\aleph_0$.

Theorem 5.9. The set of integers is denumerable.

Proof. We define $f\colon \mathbf{N} \to \mathbf{Z}$ by

$$f(x) = \begin{cases} \dfrac{x}{2} & \text{if } x \text{ is even} \\[2ex] \dfrac{1-x}{2} & \text{if } x \text{ is odd.} \end{cases}$$

Thus $f(1) = 0$, $f(2) = 1$, $f(3) = -1$, $f(4) = 2$, $f(5) = -2$, and so on. Pictorially, f is represented:

$$\begin{array}{cccccccc} \mathbf{N} = \{1, & 2, & 3, & 4, & 5, & 6, & 7, \ldots\} \\ \downarrow & \downarrow & \downarrow & \downarrow & \downarrow & \downarrow & \downarrow \\ \mathbf{Z} = \{0, & 1, & -1, & 2, & -2, & 3, & -3, \ldots\} \end{array}$$

We claim f is one-to-one. Suppose $f(x) = f(y)$. If x and y are both even, then $x/2 = y/2$, and thus $x = y$. If x and y are both odd, then $(1 - x)/2 = (1 - y)/2$, so $1 - x = 1 - y$ and $x = y$. If one of x and y is even and the other is odd, then $f(x)$ and $f(y)$ have opposite signs, so $f(x) \neq f(y)$. Thus, whenever $f(x) = f(y)$, $x = y$.

It remains to show that the function f maps onto $\mathbf{Z}$. If $w \in \mathbf{Z}$ and $w > 0$, then $2w$ is even and $f(2w) = (2w)/2 = w$. If $w \in \mathbf{Z}$ and $w \leq 0$, then $1 - 2w$ is an odd natural number and $f(1 - 2w) = [1 - (1 - 2w)]/2 = (2w)/2 = w$. In either case, if $w \in \mathbf{Z}$, then $w \in \text{Rng}(f)$. ■

We have seen two examples of infinite sets that are denumerable, but no example, as yet, of a set which is uncountable (infinite but not denumerable). Before considering the next theorem, which states that the set of real numbers in the interval (0, 1) is such an example, we need to review decimal expressions for real numbers. In its decimal form, any real number in (0, 1) may be written as $0.a_1a_2a_3a_4\ldots$, where each a_i is an integer, $0 \leq a_i \leq 9$. In this form, $\frac{7}{12} = .583333\ldots$, which is abbreviated to $.58\overline{3}$ to indicate that the 3 is repeated. A block of digits may also be repeated, as in $\frac{23}{28} = .82\overline{142857}$. The number $x = 0.a_1a_2a_3\ldots$ is said to be in **normalized form** iff there is no k such that for all $n > k$, $a_n = 9$. For example, $.82\overline{142857}$ and $\frac{2}{5} = .4\overline{0}$ are in normalized form, but $.4\overline{9}$ is not. Every real number can be expressed uniquely in normalized form. Both $.4\overline{9}$ and $.5\overline{0}$ represent the same real number $\frac{1}{2}$ but only $.5\overline{0}$ is normalized. The importance of normalizing decimals is that two decimal numbers in normalized form are equal iff they have identical digits in each decimal position.

Theorem 5.10. The interval (0, 1) is uncountable.

Proof. We must show (0, 1) is neither finite nor denumerable. The interval (0, 1) is not finite since it contains the infinite subset $\{\frac{1}{2}, \frac{1}{3}, \frac{1}{4}, \ldots\}$. ⟨*See Theorem 5.5.*⟩

Suppose there is a function $f: \mathbf{N} \to (0, 1)$ that is one-to-one. We will show that f does not map onto (0, 1). Writing the images of the elements of $\mathbf{N}$ in normalized form, we have

$$\begin{aligned} f(1) &= 0.a_{11}a_{12}a_{13}a_{14}a_{15}\ldots \\ f(2) &= 0.a_{21}a_{22}a_{23}a_{24}a_{25}\ldots \\ f(3) &= 0.a_{31}a_{32}a_{33}a_{34}a_{35}\ldots \\ f(4) &= 0.a_{41}a_{42}a_{43}a_{44}a_{45}\ldots \\ &\vdots \\ f(n) &= 0.a_{n1}a_{n2}a_{n3}a_{n4}a_{n5}\ldots \\ &\vdots \end{aligned}$$

Now let b be the number $b = 0.b_1b_2b_3b_4b_5\ldots$, where

$$b_i = \begin{cases} 5 & \text{if } a_{ii} \neq 5 \\ 3 & \text{if } a_{ii} = 5 \end{cases}$$ ⟨*The choice of 3 and 5 is arbitrary.*⟩

Then b is not the image of any $n \in \mathbf{N}$, because it differs from $f(n)$ in the nth decimal place. We conclude there is no one-to-one and onto function from $\mathbf{N}$ to (0, 1) and hence (0, 1) is not denumerable. ■

The interval (0, 1) is our first example of an uncountable set. The cardinal number of (0, 1) is defined to be **c** (which stands for continuum) and is the only infinite cardinal other than $\aleph_0$ we will mention by name.

Theorem 5.11. The set $\mathbf{R}$ is uncountable and has cardinality **c**.

Proof. Define $f: (0, 1) \to \mathbf{R}$ by $f(x) = \tan(\pi x - \pi/2)$. See figure 5.3. The function f is a contraction and translation of one branch of the tangent function and is one-to-one and onto $\mathbf{R}$. ■

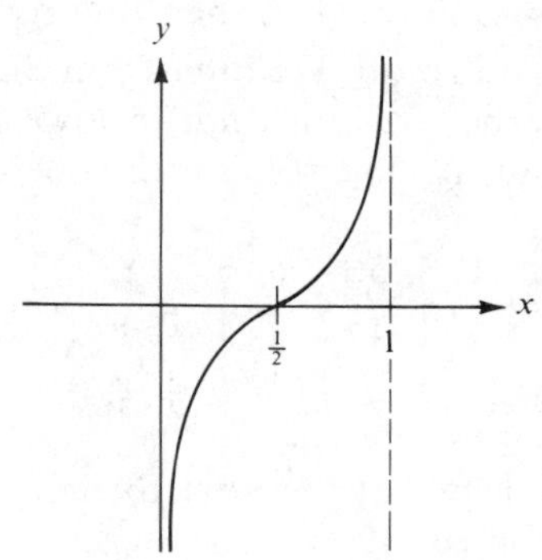

Figure 5.3

We turn now to the set of rational numbers. Since $\mathbf{N} \subseteq \mathbf{Q} \subseteq \mathbf{R}$, you should suspect that the cardinality of $\mathbf{Q}$ is $\aleph_0$, or **c**, or some infinite cardinal "in between" (a term to be made precise in the next section). You might also suspect that, since there are infinitely many rationals between any two rationals, $\mathbf{Q}$ is not denumerable, but this is not the case. Georg Cantor (1845–1918) first showed that $\mathbf{Q}^+$ (the positive rationals) is indeed denumerable through a clever rearrangement of $\mathbf{Q}^+$.

Every element in $\mathbf{Q}^+$ may be expressed as p/q for some $p, q \in \mathbf{N}$. Thus the elements of this set can be presented as in figure 5.4, where the nth row contains all the positive fractions with denominator n:

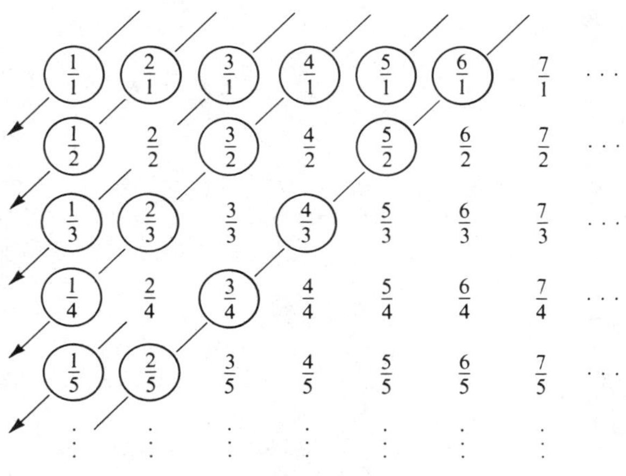

Figure 5.4

To show $\mathbf{Q}^+$ is denumerable, we list its elements in the order indicated by the diagonal arrows. First are all fractions in which the sum of the numerator and denominator is 2, (only $\frac{1}{1}$) then those whose sum is 3, ($\frac{2}{1}$, $\frac{1}{2}$), then those whose sum is 4, and so on. We omit from the list $\frac{2}{2}$ $(=\frac{1}{1})$, $\frac{4}{2}$ $(=\frac{2}{1})$, $\frac{3}{3}$ $(=\frac{2}{2}=\frac{1}{1})$, $\frac{2}{4}$ $(=\frac{1}{2})$, and all other fractions not in lowest terms. The remaining numbers are circled in the array. The result is the one-to-one correspondence

$$\begin{array}{rl}\mathbf{N} &= \{1,\ 2,\ 3,\ 4,\ 5,\ 6,\ 7,\ 8,\ 9,\ 10,\ 11,\ 12,\ 13,\ 14, \ldots\} \\ & \quad\ \downarrow\ \ \downarrow\ \ \downarrow\ \ \downarrow\ \ \downarrow\ \ \downarrow\ \ \downarrow\ \ \downarrow\ \ \downarrow\ \ \ \downarrow\ \ \ \downarrow\ \ \ \downarrow\ \ \ \downarrow\ \ \ \downarrow \\ \mathbf{Q}^+ &= \{\frac{1}{1}, \frac{2}{1}, \frac{1}{2}, \frac{3}{1}, \frac{1}{3}, \frac{4}{1}, \frac{3}{2}, \frac{2}{3}, \frac{1}{4}, \frac{5}{1}, \frac{1}{5}, \frac{6}{1}, \frac{5}{2}, \frac{4}{3}, \ldots\},\end{array}$$

which suggests a proof of the next theorem. See also the example following Theorem 5.28 on page 112.

Theorem 5.12. The set $\mathbf{Q}^+$ of positive rationals is denumerable.

Adding one or any finite number of elements to a finite set yields a finite set. Our next theorems provide analogs of these results for denumerable sets. The important distinction to be made is that adding finitely many or denumerably many elements does not change the cardinality of a denumerable set.

Theorem 5.13. If A is denumerable, then $A \cup \{x\}$ is denumerable.

Proof. Let $f: \mathbf{N} \xrightarrow[\text{onto}]{1-1} A$. If $x \in A$, $A \cup \{x\} = A$, which is denumerable. If $x \notin A$, define $g: \mathbf{N} \to A \cup \{x\}$ by

$$g(n) = \begin{cases} x & \text{if } n = 1 \\ f(n-1) & \text{if } n > 1. \end{cases}$$

It is now straightforward to verify that g is a one-to-one correspondence between $\mathbf{N}$ and $A \cup \{x\}$, which proves $A \cup \{x\}$ is denumerable. ■

Theorem 5.13 may be loosely restated as $\aleph_0 + 1 = \aleph_0$. Its proof is illustrated by the story of the Infinite Hotel.

The Infinite Hotel has $\aleph_0$ rooms and is full to capacity with one person in each room. You approach the desk clerk and ask for a room. When the clerk explains that each room is already occupied, you say, "There is room for me! For each n let the person in room n move to room $n + 1$. Then I will move into room 1, and everyone will have a room as before." There are $\aleph_0 + 1$ people and they fit exactly into the $\aleph_0$ rooms.

Rooms can also be found for any finite number of additional people (Theorem 5.14) or any denumerable number of additional people (Theorem 5.15).

Theorem 5.14. If A is denumerable and B is finite, then $A \cup B$ is denumerable.

Proof. Exercise 2. ■

Theorem 5.15. If A and B are disjoint denumerable sets, then $A \cup B$ is denumerable.

Proof. Let $f: \mathbf{N} \xrightarrow[\text{onto}]{1-1} A$ and $g: \mathbf{N} \xrightarrow[\text{onto}]{1-1} B$. Define $h: \mathbf{N} \to A \cup B$ via

$$h(n) = \begin{cases} f\left(\dfrac{n+1}{2}\right) & \text{if } n \text{ is odd} \\ g(n/2) & \text{if } n \text{ is even.} \end{cases}$$

It is left as an exercise to show that h is a one-to-one correspondence from $\mathbf{N}$ onto $A \cup B$ (exercise 3). ■

Theorems 5.13 and 5.15 provide a simple proof that the rationals are denumerable.

Theorem 5.16. The set $\mathbf{Q}$ of all rationals is denumerable.

Proof. By Theorem 5.12, $\mathbf{Q}^+$ is denumerable. By Theorem 5.13, $\mathbf{Q}^+ \cup \{0\}$ is denumerable. The function associating each rational with its negative is a one-to-one correspondence from $\mathbf{Q}^+$ to $\mathbf{Q}^-$ (the negative rationals). Thus $\mathbf{Q}^-$ is denumerable. Therefore by Theorem 5.15, $\mathbf{Q} = (\mathbf{Q}^+ \cup \{0\}) \cup \mathbf{Q}^-$ is denumerable. ■

Exercises 5.2

1. Prove that the following sets are denumerable.
   ★ (a) D^+, the odd positive integers
   (b) T^+, the positive integral multiples of 3
   (c) T, the integer multiples of 3
   (d) $\{n: n \in \mathbf{N} \text{ and } n \geq 7\}$
   ☆ (e) $\{x: x \in \mathbf{Z} \text{ and } x < -12\}$
   (f) $\mathbf{N} - \{5, 6\}$
2. Prove Theorem 5.14.
3. Prove that the function h of Theorem 5.15 is one-to-one.
4. Prove that
   (a) $(0, 1) \approx (1, 2)$ and hence conclude the cardinality of $(1, 2)$ is $\mathbf{c}$
   ★ (b) $(0, 1) \approx (4, 6)$
   (c) for any $a, b, c, d \in \mathbf{R}$ with $a < b$ and $c < d$, $(a, b) \approx (c, d)$
5. Prove that
   (a) $(3, \infty) \approx (11, \infty)$
   (b) $(-\infty, 6) \approx (4, \infty)$
   ☆ (c) $(0, 1) \approx (1, \infty)$
   (d) for any $a \in \mathbf{R}$, $(0, 1) \approx (a, \infty)$
6. What is the 28th term in the sequence of rationals given in the discussion of Theorem 5.12?
7. Prove that a set is infinite iff it contains an infinite subset.
8. Which sets have cardinal number $\aleph_0$? $\mathbf{c}$?
   ★ (a) $\mathbf{R} - \mathbf{Q}$
   (b) $(5, \infty)$
   ★ (c) $\{1/n: n \in \mathbf{N}\}$
   (d) $\{2^x: x \in \mathbf{N}\}$
   ★ (e) $\{(p, q) \in \mathbf{R} \times \mathbf{R}: p + q = 1\}$
   (f) $\{(p, q) \in \mathbf{R} \times \mathbf{R}: q = \sqrt{1 - p^2}\}$
9. (a) Show that for all $k \in \mathbf{N}$, $\mathbf{N} - \mathbf{N}_k$ is denumerable.
   ★ (b) Show that if $S \subseteq \mathbf{N}$ and S is finite, then $\mathbf{N} - S$ is denumerable.
   (c) Show that if $B \subseteq A$, A is denumerable, and B is finite, then $A - B$ is denumerable.
10. ☆ Let S be the set of all sequences of 0's and 1's. For example, 1010101... and 010110111... are in S. Prove that S is uncountable.
11. Give another proof of Theorem 5.11 by showing that $f(x) = (x - 1/2)/[x(x - 1)]$ is a one-to-one correspondence from $(0, 1)$ onto $\mathbf{R}$.
12. ☆ Let $\{A_i: i = 1, 2, \ldots, n\}$ be a finite pairwise disjoint collection of denumerable sets. Prove $\bigcup_{i=1}^{n} A_i$ is denumerable.
13. Give an example of two distinct infinite sets A and B such that
   (a) $A - B$ is empty
   (b) $A - B$ is finite and nonempty
   (c) $A - B$ is denumerable
   (d) $A - B$ is uncountable
14. **Proofs to Grade.**
   ★ (a) **Claim.** The sets E^+ of even natural numbers and D^+ of odd natural numbers are equivalent.
   "Proof." E^+ is an infinite subset of $\mathbf{N}$. Thus E^+ is denumerable. Likewise D^+ is an infinite subset of $\mathbf{N}$ and is denumerable. Therefore, $E^+ \approx D^+$. ■
   (b) **Claim.** If A is infinite and $x \notin A$, then $A \cup \{x\}$ is infinite.
   "Proof." Let A be infinite. Then $A \approx \mathbf{N}$. Let $f: \mathbf{N} \to A$ be a one-to-one correspondence. Then $g: \mathbf{N} \to A \cup \{x\}$, defined by

$$g(t) = \begin{cases} x & \text{if } t = 1 \\ f(t-1) & \text{if } t > 1, \end{cases}$$

   is one-to-one and onto $A \cup \{x\}$. Thus $\mathbf{N} \approx A \cup \{x\}$, so $A \cup \{x\}$ is infinite. ■

★ (c) **Claim.** If $A \cup B$ is infinite, then A and B are infinite.
"Proof." This is a proof by contrapositive so assume the denial of the consequent. Thus, assume A and B are finite. Then, by Theorem 5.6, $A \cup B$ is finite, which is a denial of the antecedent. Therefore, the result is proved. ■

★ (d) **Claim.** If a set A is infinite, then A is equivalent to a proper subset of A.
"Proof." Let $A = \{x_1, x_2, \ldots\}$. Choose $B = \{x_2, x_3, \ldots\}$. Then B is a proper subset of A. The function $f: A \to B$ defined by $f(x_k) = x_{k+1}$ is clearly one-to-one and onto B. Thus $A \approx B$. ■

(e) **Claim.** (Theorem 5.12) The set $\mathbb{Q}^+$ of positive rationals is denumerable.
"Proof." Consider the positive rationals in the array on page 100. Consider the order formed by listing all the rationals in the first row, then the second row, and so forth. Omitting fractions that are not in lowest terms, we have an ordering of $\mathbb{Q}^+$ in which every positive rational appears. Therefore $\mathbb{Q}^+$ is denumerable. ■

(f) **Claim.** If A and B are infinite, then $A \approx B$.
"Proof." Suppose A and B are infinite sets. Let $A = \{a_1, a_2, a_3, \ldots\}$ and $B = \{b_1, b_2, b_3, \ldots\}$. Define $f: A \to B$ as in the picture

$$\begin{array}{c} \{a_1, a_2, a_3, a_4, \ldots\} \\ \downarrow\ \ \downarrow\ \ \downarrow\ \ \downarrow \\ \{b_1, b_2, b_3, b_4, \ldots\} \end{array}$$

Then since we never run out of elements in either set, f is one-to-one and onto B, so $A \approx B$. ■

SECTION 5.3. THE ORDERING OF CARDINAL NUMBERS

The theory of infinite sets was developed by Georg Cantor, primarily in papers appearing in 1895 and 1897. He described a cardinal number of a set M as "the general concept which, with the aid of our intelligence, results from M when we abstract from the nature of its various elements and from the order of their being given." This definition has been criticized as being less precise and more mystical than a definition in mathematics ought to be. Other definitions were given, and eventually the concept of cardinal number was made precise. One way this may be done is to determine a fixed set from each equivalence class of sets under the relation $\approx$, and then to call this set the cardinal number of each set in the class. Under such a procedure we would think of the number 0 as *being* the empty set; the number 1 as being the set whose only element is the number 0. That is, $1 = \{0\}$; $2 = \{0, 1\}$; $3 = \{0, 1, 2\}$, and so on.

We will not be concerned with a precise definition of a cardinal number. For our purposes the essential point is that the **cardinal number** $\overline{\overline{A}}$ of a set A is an object associated with all sets equivalent to A and with no other set. The double bars on $\overline{\overline{A}}$ are suggestive of the double abstraction referred to by Cantor. For example, $\overline{\overline{\{5\}}} = \overline{\overline{\{1\}}} = 1$ and $\overline{\overline{\{p, q, r, s\}}} = \overline{\overline{\{1, 2, 3, 4\}}} = 4$.

Cardinal numbers may be ordered (compared) in the following manner:

Definitions.

(a) $\overline{\overline{A}} = \overline{\overline{B}}$ if and only if $A \approx B$; otherwise $\overline{\overline{A}} \neq \overline{\overline{B}}$.
(b) $\overline{\overline{A}} \leq \overline{\overline{B}}$ if and only if there exists $f: A \xrightarrow{1-1} B$.
(c) $\overline{\overline{A}} < \overline{\overline{B}}$ if and only if $\overline{\overline{A}} \leq \overline{\overline{B}}$ and $\overline{\overline{A}} \neq \overline{\overline{B}}$.

$\overline{\overline{A}} < \overline{\overline{B}}$ is read "the cardinality of A is strictly less than the cardinality of B" while $\leq$ is read "less than or equal to." In addition, we use $\overline{\overline{A}} \not< \overline{\overline{B}}$ and $\overline{\overline{A}} \not\leq \overline{\overline{B}}$ to denote the denials of $\overline{\overline{A}} < \overline{\overline{B}}$ and $\overline{\overline{A}} \leq \overline{\overline{B}}$, respectively. In general, a proof of $\overline{\overline{A}} \leq \overline{\overline{B}}$ will involve constructing a one-to-one function, while a proof of $\overline{\overline{A}} < \overline{\overline{B}}$ will require a proof of $\overline{\overline{A}} \leq \overline{\overline{B}}$ together with a proof of $\overline{\overline{A}} \neq \overline{\overline{B}}$ by contradiction.

Since 1, 2, 3, ... are cardinal numbers, the natural numbers may be viewed as a subset of the collection of all cardinal numbers. In this sense the ordering theorems that follow may be interpreted as extensions of those for $\mathbb{N}$. Those not proved are (as usual) left as exercises at the end of this section (exercise 6).

Theorem 5.17. For sets A, B, and C,

(a) (Reflexivity) $\overline{\overline{A}} \leq \overline{\overline{A}}$.
(b) (Transitivity of $=$) If $\overline{\overline{A}} = \overline{\overline{B}}$ and $\overline{\overline{B}} = \overline{\overline{C}}$, then $\overline{\overline{A}} = \overline{\overline{C}}$.
(c) (Transitivity of $\leq$) If $\overline{\overline{A}} \leq \overline{\overline{B}}$ and $\overline{\overline{B}} \leq \overline{\overline{C}}$, then $\overline{\overline{A}} \leq \overline{\overline{C}}$.
(d) $\overline{\overline{A}} \leq \overline{\overline{B}}$ iff $\overline{\overline{A}} < \overline{\overline{B}}$ or $\overline{\overline{A}} = \overline{\overline{B}}$.
(e) If $A \subseteq B$, then $\overline{\overline{A}} \leq \overline{\overline{B}}$.
(f) $\overline{\overline{A}} \leq \overline{\overline{B}}$ iff there is a subset W of B such that $\overline{\overline{W}} = \overline{\overline{A}}$.

Proof.

(c) Suppose $\overline{\overline{A}} \leq \overline{\overline{B}}$ and $\overline{\overline{B}} \leq \overline{\overline{C}}$. Then there exist functions $f: A \xrightarrow{1-1} B$ and $g: B \xrightarrow{1-1} C$. Since the composite $g \circ f: A \to C$ is one-to-one, we conclude $\overline{\overline{A}} \leq \overline{\overline{C}}$.

(e) Let $A \subseteq B$. We note that the inclusion map $i: A \to B$, given by $i(a) = a$, is one-to-one, whence $\overline{\overline{A}} \leq \overline{\overline{B}}$. ■

Although Cantor developed many aspects of the theory of infinite sets, his name remains attached particularly to the next theorem, which states that the cardinality of a set A is strictly less than the cardinality of its power set. We already know the result to be true if A is a finite set with n elements, since $\mathscr{P}(A)$ has 2^n elements. The proof given here holds for all sets A. An immediate consequence is that there can be no largest cardinal number (see exercise 7).

Theorem 5.18. (Cantor's Theorem) For every set A, $\overline{\overline{A}} < \overline{\overline{\mathscr{P}(A)}}$.

Proof. To show $\overline{\overline{A}} < \overline{\overline{\mathscr{P}(A)}}$, we must show that (i) $\overline{\overline{A}} \leq \overline{\overline{\mathscr{P}(A)}}$, and (ii) $\overline{\overline{A}} \neq \overline{\overline{\mathscr{P}(A)}}$. Part (i) follows from the fact that $F\colon A \to \mathscr{P}(A)$ defined by $F(x) = \{x\}$ is one-to-one.

To prove (ii), suppose $\overline{\overline{A}} = \overline{\overline{\mathscr{P}(A)}}$; that is, assume $A \approx \mathscr{P}(A)$. Then there exists $g\colon A \xrightarrow[\text{onto}]{1-1} \mathscr{P}(A)$. Let $B = \{y \in A : y \notin g(y)\}$. Since $B \subseteq A$, $B \in \mathscr{P}(A)$, and since g is onto $\mathscr{P}(A)$, $B = g(z)$ for some $z \in A$. Now either $z \in B$ or $z \notin B$. If $z \in B$, then $z \notin g(z) = B$, a contradiction. Similarly, $z \notin B$ implies $z \in g(z)$, which implies $z \in B$, again a contradiction. We conclude $A \not\approx \mathscr{P}(A)$ and hence $\overline{\overline{A}} < \overline{\overline{\mathscr{P}(A)}}$. ■

It appears to be obvious that if B has at least as many elements as A ($\overline{\overline{A}} \leq \overline{\overline{B}}$) and A has at least as many elements as B ($\overline{\overline{B}} \leq \overline{\overline{A}}$), then A and B are equivalent ($\overline{\overline{A}} = \overline{\overline{B}}$). The proof, however, is not obvious. The situation may be represented as in figure 5.5. From $\overline{\overline{A}} \leq \overline{\overline{B}}$ and $\overline{\overline{B}} \leq \overline{\overline{A}}$, there are functions $F\colon A \xrightarrow{1-1} B$ and $G\colon B \xrightarrow{1-1} A$. The problem is to construct $H\colon A \to B$, which is both one-to-one and onto B. Cantor (1895) solved this problem, but his proof used the controversial Axiom of Choice (to be discussed in the next section). Proofs not depending on the Axiom of Choice were given independently by Ernest Schröder in 1896 and two years later by Felix Bernstein.

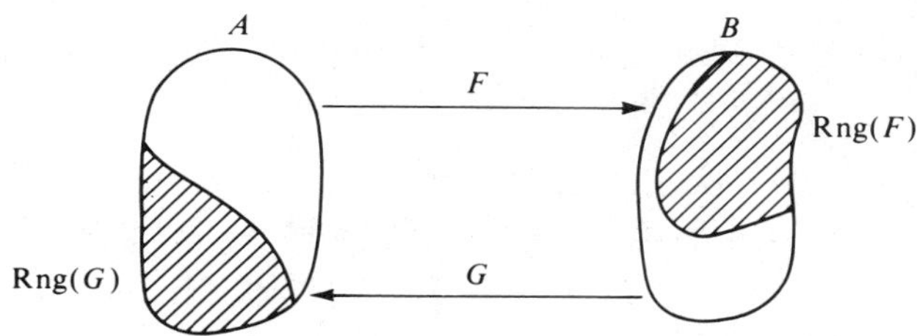

Figure 5.5

Theorem 5.19. (Cantor-Schröder-Bernstein Theorem).
If $\overline{\overline{A}} \leq \overline{\overline{B}}$ and $\overline{\overline{B}} \leq \overline{\overline{A}}$, then $\overline{\overline{A}} = \overline{\overline{B}}$.

Proof. We may assume that A and B are disjoint, for otherwise we could replace A and B with the equivalent sets $A \times \{0\}$ and $B \times \{1\}$, respectively. Let $F\colon A \xrightarrow{1-1} B$, with $D = \operatorname{Rng}(F)$, and let $G\colon B \xrightarrow{1-1} A$, with $C = \operatorname{Rng}(G)$. Define a **string** to be a function $f\colon \mathbf{N} \to A \cup B$ such that

$$f(1) \in B - D,$$
$$f(n) \in B \text{ implies } f(n+1) = G(f(n)),$$

and

$$f(n) \in A \text{ implies } f(n+1) = F(f(n)).$$

We think of a string as a sequence of elements of $A \cup B$ with first term in $B - D$, and such that the terms are alternately in B and in A. Each element of $B - D$ is the first term of some string.

Let $W = \{x \in A : x \text{ is a term of some string}\}$ and define $H: A \to B$ by

$$H(x) = \begin{cases} F(x) & \text{if } x \in A - W \\ G^{-1}(x) & \text{if } x \in W \end{cases}$$

We will show that $H: A \xrightarrow[\text{onto}]{H} B$. Because F and G^{-1} are one-to-one, to show H is one-to-one we need only consider the possibility that $F(a) = G^{-1}(b)$ for some $a \in A - W$ and $b \in W$. Suppose $b = G(F(a))$ and b is in a string f. Then $b = f(2n)$ for some $n \geq 1$ and $F(a) = f(2n-1)$. If $n = 1$, then $F(a) = f(1)$. But $f(1) \notin Rng(F)$. Thus $n \geq 2$, so $a = f(2n-2)$. This is a contradiction, since $a \in A - W$.

We next show that H is onto B. Let $b \in B$. We must show that for some $x \in A$, $H(x) = b$.

Case 1. If $G(b) \in W$, then $H(G(b)) = G^{-1}(G(b)) = b$. Hence, in this case, we may take $x = G(b)$.

Case 2. If $G(b) \notin W$, then there exists $x \in A$ such that $F(x) = b$. ⟨*For if there were no such x, then* $b \in B - D$, *and so some string would begin at b, and* $G(b)$ *would be a term of that string, hence in W.*⟩ Furthermore, $x \in A - W$, for if x were a term of some string, then also $F(x)$ and $G(F(x)) = G(b)$ would be terms of the same string, which would imply $G(b) \in W$. Since $x \in A - W$, $H(x) = F(x) = b$. ■

The Cantor-Schröder-Bernstein Theorem leads to several interesting extensions of Theorem 5.17. Parts (a) and (c) are proved here; (b) and (d) are given as exercise 11.

Corollary 5.20. For sets A, B, and C,

(a) if $\overline{\overline{A}} \leq \overline{\overline{B}}$, then $\overline{\overline{B}} \not< \overline{\overline{A}}$.
(b) if $\overline{\overline{A}} \leq \overline{\overline{B}}$ and $\overline{\overline{B}} < \overline{\overline{C}}$, then $\overline{\overline{A}} < \overline{\overline{C}}$.
(c) if $\overline{\overline{A}} < \overline{\overline{B}}$ and $\overline{\overline{B}} \leq \overline{\overline{C}}$, then $\overline{\overline{A}} < \overline{\overline{C}}$.
(d) if $\overline{\overline{A}} < \overline{\overline{B}}$ and $\overline{\overline{B}} < \overline{\overline{C}}$, then $\overline{\overline{A}} < \overline{\overline{C}}$.

Proof.

(a) Suppose $\overline{\overline{B}} < \overline{\overline{A}}$. Then $\overline{\overline{A}} \neq \overline{\overline{B}}$ and $\overline{\overline{B}} \leq \overline{\overline{A}}$. Combining this with the hypothesis that $\overline{\overline{A}} \leq \overline{\overline{B}}$, we conclude by the Cantor-Schröder-Bernstein Theorem that $\overline{\overline{A}} = \overline{\overline{B}}$, which is a contradiction. Therefore, $\overline{\overline{B}} \not< \overline{\overline{A}}$.
(c) Suppose $\overline{\overline{A}} < \overline{\overline{B}}$ and $\overline{\overline{B}} \leq \overline{\overline{C}}$. Then $\overline{\overline{A}} \leq \overline{\overline{B}}$; so by Theorem 5.15 (b), $\overline{\overline{A}} \leq \overline{\overline{C}}$. Suppose $\overline{\overline{A}} \not< \overline{\overline{C}}$. Then $\overline{\overline{A}} = \overline{\overline{C}}$, which implies $\overline{\overline{C}} \leq \overline{\overline{A}}$. But $\overline{\overline{B}} \leq \overline{\overline{C}}$ and $\overline{\overline{C}} \leq \overline{\overline{A}}$ imply $\overline{\overline{B}} \leq \overline{\overline{A}}$. Combining this with $\overline{\overline{A}} \leq \overline{\overline{B}}$, we conclude by the Cantor-Schröder-Bernstein Theorem that $\overline{\overline{A}} = \overline{\overline{B}}$. Since this contradicts $\overline{\overline{A}} < \overline{\overline{B}}$, we have $\overline{\overline{A}} < \overline{\overline{C}}$. ■

It is tempting to extend our results even further to include the converse of Corollary 5.20 (a): "If $\overline{\overline{A}} \nleq \overline{\overline{B}}$, then $\overline{\overline{B}} \leq \overline{\overline{A}}$." (As far as we know now, for two given sets A and B, both $\overline{\overline{A}} \leq \overline{\overline{B}}$ and $\overline{\overline{B}} \leq \overline{\overline{A}}$ may be false.) The Cantor-Schröder-Bernstein Theorem turned out to be more difficult to prove than one would have guessed from its simple statement, but the situation regarding the converse of Corollary 5.20 (a) is even more remarkable. This is discussed in the next section, where "If $\overline{\overline{A}} \nleq \overline{\overline{B}}$, then $\overline{\overline{B}} \leq \overline{\overline{A}}$" is rephrased as "Either $\overline{\overline{A}} < \overline{\overline{B}}$ or $\overline{\overline{A}} = \overline{\overline{B}}$ or $\overline{\overline{B}} < \overline{\overline{A}}$."

Exercises 5.3

★ 1. Prove that if $n \in \mathbf{N}$, then $n < \aleph_0$.

2. Prove $\aleph_0 < \mathbf{c}$.

3. Prove that if $\overline{\overline{A}} \leq \overline{\overline{B}}$ and $\overline{\overline{B}} = \overline{\overline{C}}$, then $\overline{\overline{A}} \leq \overline{\overline{C}}$.

4. Prove that if $\overline{\overline{A}} \leq \overline{\overline{B}}$ and $\overline{\overline{A}} = \overline{\overline{C}}$, then $\overline{\overline{C}} \leq \overline{\overline{B}}$.

5. Arrange the following cardinal numbers in order:

★ (a) $\overline{\overline{(0, 1)}}$, $\overline{\overline{[0, 1]}}$, $\overline{\overline{\{0, 1\}}}$, $\overline{\overline{\{0\}}}$, $\overline{\overline{\mathscr{P}(\mathbf{R})}}$, $\overline{\overline{\mathbf{Q}}}$, $\overline{\overline{\varnothing}}$, $\overline{\overline{\mathbf{R} - \mathbf{Q}}}$, $\overline{\overline{\mathscr{P}(\mathscr{P}(\mathbf{R}))}}$, $\overline{\overline{\mathbf{R}}}$

(b) $\overline{\overline{\mathbf{Q} \cup \{\pi\}}}$, $\overline{\overline{\mathbf{R} - \{\pi\}}}$, $\overline{\overline{\mathscr{P}(\{0, 1\})}}$, $\overline{\overline{[0, 2]}}$, $\overline{\overline{(0, \infty)}}$, $\overline{\overline{\mathbf{Z}}}$, $\overline{\overline{\mathbf{R} - \mathbf{Z}}}$, $\overline{\overline{\mathscr{P}(\mathbf{R})}}$

6. Prove the remaining parts of Theorem 5.17.

☆ 7. Prove that there is no largest cardinal number.

8. Apply the proof of the Cantor-Schröder-Bernstein Theorem to this situation: $A = \{2, 3, 4, 5, \ldots\}$, $B = \{\frac{1}{2}, \frac{1}{3}, \frac{1}{4}, \ldots\}$, $F\colon A \to B$ where $F(x) = \dfrac{1}{x + 6}$, and $G\colon B \to A$ where $G(x) = \dfrac{1}{x} + 5$. Note that $\frac{1}{3}$ and $\frac{1}{4}$ are in $B - \operatorname{Rng}(F)$. Let f be the string which begins at $\frac{1}{3}$, and g be the string which begins at $\frac{1}{4}$.
   (a) Find $f(1), f(2), f(3), f(4)$.
   (b) Find $g(1), g(2), g(3), g(4)$.
   (c) Define H as in the proof of the Cantor-Schröder-Bernstein Theorem and find $H(2)$, $H(8)$, $H(13)$, and $H(20)$.

9. Suppose there exist functions $f\colon A \xrightarrow{1-1} B$, $g\colon B \xrightarrow{1-1} C$, and $h\colon C \xrightarrow{1-1} A$. Prove $A \approx B \approx C$. Do not assume the functions map onto their codomains.

10. If possible, give an example of
   (a) functions f and g such that $f\colon \mathbf{Q} \xrightarrow{1-1} \mathbf{N}$, $g\colon \mathbf{N} \xrightarrow{1-1} \mathbf{Q}$, but neither f nor g is an onto map.
   ★ (b) a function $f\colon \mathbf{R} \xrightarrow{1-1} \mathbf{N}$.
   (c) a function $f\colon \mathscr{P}(\mathbf{N}) \xrightarrow{1-1} \mathbf{N}$.

11. Prove parts (b) and (d) of Corollary 5.20.

12. Prove that there is no universal set U of all sets.

13. **Proofs to Grade.**
   (a) **Claim.** If $\overline{\overline{A}} \leq \overline{\overline{B}}$ and $\overline{\overline{A}} = \overline{\overline{C}}$, then $\overline{\overline{C}} \leq \overline{\overline{B}}$.
   "Proof." Assume $\overline{\overline{A}} \leq \overline{\overline{B}}$ and $\overline{\overline{A}} = \overline{\overline{C}}$. Then there exists a function f such that $f\colon A \xrightarrow{1-1} B$. Since $\overline{\overline{A}} = \overline{\overline{C}}$, $f\colon C \xrightarrow{1-1} B$. Therefore, $\overline{\overline{C}} \leq \overline{\overline{B}}$. ■
   ★ (b) **Claim.** If $B \subseteq C$ and $\overline{\overline{B}} = \overline{\overline{C}}$, then $B = C$.
   "Proof." Suppose $B \neq C$. Then B is a proper subset of C. Thus $C - B \neq \varnothing$. This implies $\overline{\overline{C - B}} > 0$. But $C = B \cup (C - B)$ and, since B and $C - B$ are disjoint, $\overline{\overline{C}} = \overline{\overline{B}} + \overline{\overline{(C - B)}}$. By hypothesis, $\overline{\overline{B}} = \overline{\overline{C}}$. Thus $\overline{\overline{(C - B)}} = 0$, a contradiction. ■

(c) **Claim.** If $\overline{\overline{A}} \leq \overline{\overline{B}}$ and $\overline{\overline{B}} = \overline{\overline{C}}$, then $\overline{\overline{A}} \leq \overline{\overline{C}}$.

"Proof." Assume $\overline{\overline{A}} \leq \overline{\overline{B}}$. Then, since $\overline{\overline{B}} = \overline{\overline{C}}$, we have $\overline{\overline{A}} \leq \overline{\overline{C}}$ by substitution. ■

★ (d) **Claim.** If $A \neq \varnothing$ and $\overline{\overline{A}} \leq \overline{\overline{B}}$, then there exists a function $f\colon B \xrightarrow{\text{onto}} A$.

"Proof." From $\overline{\overline{A}} \leq \overline{\overline{B}}$, we know there exists $g\colon A \xrightarrow{1-1} B$. Fix a particular $a^* \in A$. Define $f\colon B \to A$ as follows:

(1) For $b \in B$, if $b \in \text{Rng}(g)$, then $g^{-1}(\{b\})$ consists of a single element of A, since g is one-to-one. Define $f(b)$ to be equal to that element of A.

(2) If $b \notin \text{Rng}(g)$, define $f(b) = a^*$.

The function f is onto A, for if $a \in A$, let $g(a) = b$. Then $b \in B$ and $g^{-1}(\{b\}) = a$, so $f(b) = a$. ■

SECTION 5.4. COMPARABILITY OF CARDINALS AND THE AXIOM OF CHOICE

The relations $\leq$ and $<$ on the natural numbers have several familiar properties such as reflexivity, transitivity, and so on. One goal of the last section was to establish many similar results for the relations $\leq$ and $<$ on the class of all cardinal numbers.

A useful fact about $\mathbf{N}$ is the trichotomy property: if m and n are any two natural numbers, then $m > n$, $m = n$, or $n < m$. The analog for cardinal numbers is stated in the following.

Theorem 5.21. (The Comparability Theorem). If A and B are any two sets, then $\overline{\overline{A}} < \overline{\overline{B}}$, $\overline{\overline{A}} = \overline{\overline{B}}$, or $\overline{\overline{B}} < \overline{\overline{A}}$.

Surprisingly, all our knowledge of sets and functions is insufficient to prove the Comparability Theorem. On the other hand, it is also insufficient to prove that the Comparability Theorem is false. Theorem 5.21 is an undecidable sentence.

The solution to our problem is that we may either assume Theorem 5.21 to be true (or assume true some other statement that implies comparability) or else assume the truth of some statement from which we can show comparability is false. It has become standard practice by most mathematicians to assume the Comparability Theorem is true by assuming the truth of the following statement:

The Axiom of Choice. If $\mathscr{A}$ is any collection of nonempty sets, then there exists a function F (called a choice function) from $\mathscr{A}$ to $\bigcup_{A \in \mathscr{A}} A$ such that for every $A \in \mathscr{A}$, $F(A) \in A$.

The Axiom of Choice at first appears to have little significance: from a collection of nonempty sets we can choose an element from each set. If the collection is finite, then this axiom is not needed to prove the existence of a choice function. It is only for infinite collections of sets that the result is not obvious and for which the Axiom of Choice is independent of other axioms of set theory.

Many examples and uses of the Axiom of Choice require a sophisticated background in mathematics. The example we present is not mathematical in content yet has become part of mathematical folklore.

A shoe store has in the stockroom an infinite number of pairs of shoes and an infinite number of pairs of socks. A customer asks to see one shoe of each pair and one sock of each pair. The clerk does not need the Axiom of Choice to select a shoe from each pair of shoes. His choice rule might be "From each pair of shoes, choose the left shoe." However, since the socks of any pair of socks are indistinguishable, and since there are an infinite number of pairs of socks, the clerk must employ the Axiom of Choice to show the customer one sock from each pair.

With the Axiom of Choice, we could, but will not, prove Theorem 5.21. For a proof, see R. L. Wilder, *Introduction to the Foundations of Mathematics,* 2nd Ed. (John Wiley & Sons, Inc., 1965).

Many other important theorems, in many areas of mathematics, cannot be proved without the use of the Axiom of Choice.* In fact, several crucial results are equivalent to it. Some of the consequences of the axiom are not so natural as the Comparability Theorem however, and some of them are extremely difficult to believe. (One of these is that the real numbers can be rearranged in such a way that every nonempty subset of $\mathbf{R}$ has a first element—in other words, that the reals can be well ordered.) The Axiom of Choice has been objected to because of such consequences, and also because of a lack of precision in the statement of the axiom, which does not provide any hint of a rule for constructing the choice function F. Because of these objections it is common practice to call attention to the fact that the Axiom of Choice has been used in a proof, so that anyone who is interested can attempt to find an alternate proof which does not use the axiom.

We present two theorems whose proofs require the Axiom of Choice. The first is that if there is a function f from A onto B, then A must have at least as many elements as B. The proof chooses for each $b \in B$, an $a \in A$ such that $f(a) = b$. The second theorem is that every infinite set has a denumerable subset. The axiom is used to define the denumerable subset inductively.

Theorem 5.22. If there exists a function from A onto B, then $\overline{\overline{B}} \leq \overline{\overline{A}}$.

Proof. If $B = \varnothing$, then $B \subseteq A$, so $\overline{\overline{B}} \leq \overline{\overline{A}}$. Let $f: A \to B$ be onto B where $B \neq \varnothing$. To show $\overline{\overline{B}} \leq \overline{\overline{A}}$, we must construct a function $h: B \to A$ which is one-to-one. Since f is onto B, for each $b \in B$, $b \in \operatorname{Rng}(f)$ or, equivalently, $f^{-1}(\{b\}) \neq \varnothing$. Thus $\mathscr{A} = \{f^{-1}(\{b\}): b \in B\}$ is a nonempty collection of nonempty sets. By the Axiom of Choice there exists a function g from $\mathscr{A}$ to $\bigcup_{b \in B} f^{-1}(\{b\}) = A$ such that $g(f^{-1}(\{b\})) \in f^{-1}(\{b\})$ for all $b \in B$.

*See H. Rubin and J. E. Rubin, *Equivalents of the Axiom of Choice* (North Holland Pub. Co., New Amsterdam, 1963).

Define $h: B \to A$ by $h(b) = g(f^{-1}(\{b\}))$. It remains to prove h is one-to-one. Suppose $h(b_1) = h(b_2)$. Then $g(f^{-1}(\{b_1\})) = g(f^{-1}(\{b_2\}))$. By definition of g, this element is in both $f^{-1}(\{b_1\})$ and $f^{-1}(\{b_2\})$. Thus $f^{-1}(\{b_1\}) \cap f^{-1}(\{b_2\}) \neq \varnothing$. For any $x \in f^{-1}(\{b_1\}) \cap f^{-1}(\{b_2\})$, $f(x) = b_1$, and $f(x) = b_2$. Therefore, $b_1 = b_2$. This proves h is one-to-one and that $\overline{\overline{B}} \leq \overline{\overline{A}}$. ■

Theorem 5.23. Every infinite set A has a denumerable subset.

Proof. We shall inductively define a denumerable subset of elements of A. First, since A is infinite, $A \neq \varnothing$. Choose $a_1 \in A$. Then $A - \{a_1\}$ is infinite, hence nonempty. Choose $a_2 \in A - \{a_1\}$. Note that $a_2 \neq a_1$ and $a_2 \in A$. Continuing in this fashion, suppose $a_1, \ldots, a_k$ have been defined. Then $A - \{a_1, \ldots, a_k\} \neq \varnothing$, so select any a_{k+1} from this set. By the Axiom of Choice, a_n is defined for all $n \in \mathbb{N}$. The a_n have been constructed so that each $a_n \in A$ and $a_i \neq a_j$ for $i \neq j$. Thus $B = \{a_n : n \in \mathbb{N}\}$ is a subset of A, and the function f given by $f(n) = a_n$ is a one-to-one correspondence from $\mathbb{N}$ to B. Thus B is denumerable. ■

Exercises 5.4

1. Prove that if $f: A \to B$, then $\overline{\overline{\mathrm{Rng}(f)}} \leq \overline{\overline{A}}$.
2. ★ Suppose A is a denumerable set and B is an infinite subset of A. Prove $A \approx B$.
3. Suppose $\overline{\overline{B}} < \overline{\overline{C}}$ and $\overline{\overline{B}} \not\leq \overline{\overline{A}}$. Prove $\overline{\overline{A}} < \overline{\overline{C}}$.
4. Let $\{A_i : i \in \mathbb{N}\}$ be a collection of pairwise disjoint nonempty sets. Prove that $\bigcup_{i \in \mathbb{N}} A_i$ contains a denumerable subset.
5. **Proofs to Grade.**
   - ★ (a) **Claim.** There exists an infinite set of irrational numbers, no two of which differ by a rational number.
     "Proof." For $x, y \in \mathbb{R}$, define the relation S by $x\,S\,y$ iff $x - y \in \mathbb{Q}$. It is easy to show that S is an equivalence relation. For the family of equivalence classes $\{x/S : x \in \mathbb{R}\}$, choose one element from each equivalence class except the equivalence class $\mathbb{Q}$. The set of all such chosen elements is an infinite set of irrational numbers, no two of which differ by a rational. (You may accept, without proof, the claim that this set is infinite.) ■
   - (b) **Claim.** Every infinite set A has a denumerable subset.
     "Proof." Suppose no subset of A is denumerable. Then all subsets of A must be finite. In particular $A \subseteq A$. Thus A is finite, contradicting the assumption. ■
   - ★ (c) **Claim.** Every infinite set A has a denumerable subset B.
     "Proof." If A is denumerable, let $B = A$, and we are done. Otherwise, A is uncountable. Choose $x_1 \in A$. If $A - \{x_1\}$ is denumerable, let $B = A - \{x_1\}$. Otherwise, choose $x_2 \in A - \{x_1\}$. If $A - \{x_1, x_2\}$ is denumerable, let $B = A - \{x_1, x_2\}$. Continuing in this manner, using the Axiom of Choice, we obtain a subset $C = \{x_1, x_2, \ldots\}$ such that $B = A - C$ is denumerable. ■
   - (d) **Claim.** Every infinite set has two disjoint denumerable subsets.
     "Proof." Let A be an infinite set. By Theorem 5.23, A has a denumerable subset, B. Then $A - B$ is infinite, because A is infinite and is disjoint from B. By Theorem 5.23 $A - B$ has a denumerable subset, C. Then B and C are disjoint denumerable subsets of A. ■

(e) **Claim.** Every infinite set has two disjoint denumerable subsets.
"Proof." Let A be an infinite set. By Theorem 5.23, A has a denumerable subset B. Since B is denumerable, there is a function $f: \mathbf{N} \xrightarrow[\text{onto}]{1-1} B$. Let $C = \{f(2n): n \in \mathbf{N}\}$ and $D = \{f(2n-1): n \in \mathbf{N}\}$. Then $C = \{f(2), f(4), f(6), \ldots\}$ and $D = \{f(1), f(3), f(5), \ldots\}$ are disjoint denumerable subsets of A. ■

SECTION 5.5. COUNTABLE SETS

Countable sets have been defined as those sets that are finite or denumerable. Such sets are countable in the sense that they can be "counted" by using subsets of $\mathbf{N}$: a finite set is counted by using exactly the elements of some $\mathbf{N}_k$, while a denumerable set may be counted by using exactly the elements of $\mathbf{N}$. This section will survey some of the important results on countability.

Theorem 5.24. Every subset of a countable set is countable.

Proof. Let A be a countable set and let $B \subseteq A$. If B is finite, then B is countable by definition. If B is infinite, since $B \subseteq A$, A is infinite. Thus A is denumerable. By Theorem 5.23, B has a denumerable subset C. Thus $C \subseteq B \subseteq A$, which implies $\aleph_0 = \overline{\overline{C}} \leq \overline{\overline{B}} \leq \overline{\overline{A}} = \aleph_0$. Therefore, $\overline{\overline{A}} = \overline{\overline{B}} = \aleph_0$. Thus B is denumerable and hence countable. ■

We have seen that the set $\mathbf{Q}$ of all rational numbers is countable. Thus such subsets as $\{1/n: n \in \mathbf{N}\}$, $\mathbf{Q} \cap (0, 1)$, $\{\frac{5}{6}, \frac{6}{5}, \frac{3}{7}\}$, and $\mathbf{Z}$ are countable sets.

Corollary 5.25. Let A be a set.

(a) A is countable iff A is equivalent to a subset of $\mathbf{N}$.
(b) A is countable iff $\overline{\overline{A}} \leq \aleph_0$.

Proof. Exercise 2. ■

Theorem 5.26. If A and B are countable sets, then $A \cup B$ is countable.

Proof. This theorem has been proved in the cases where A and B are finite (Corollary 5.7), where one set is denumerable and the other is finite (Theorem 5.14), and where A and B are denumerable and disjoint (Theorem 5.15). The only remaining case is where A and B are denumerable and not disjoint. Write $A \cup B$ as $A \cup (B - A)$, a union of disjoint sets. Since $B - A \subseteq B$, $B - A$ is either finite or denumerable by Theorem 5.24. If $B - A$ is finite, then $A \cup B$ is denumerable by Theorem 5.14 and, if $B - A$ is denumerable, then $A \cup B$ is denumerable by Theorem 5.15. ■

Theorem 5.26 may be extended (by induction) to any finite union of countable sets and finally to a countable union of countable sets.

Theorem 5.27. Let $\mathscr{A}$ be a finite collection of countable sets. Then $\bigcup_{A \in \mathscr{A}} A$ is countable.

Proof. Exercise 3. ■

Theorem 5.28. Let $\mathscr{A}$ be a denumerable collection of denumerable sets. Then $\bigcup_{A \in \mathscr{A}} A$ is denumerable.

Proof. Since $\mathscr{A}$ is denumerable, we may write $\mathscr{A}$ as $\{A_i : i \in \mathbf{N}\}$. We first consider the case where $\mathscr{A}$ is a pairwise disjoint collection. Since each A_i is denumerable, for each i there exists a function $f_i : \mathbf{N} \xrightarrow[\text{onto}]{1-1} A_i$. Let $f : \mathbf{N} \to \bigcup_{A \in \mathscr{A}} A$ be defined by $f(n) = f_1(n)$. Since f_1 is one-to-one, f is one-to-one and, therefore, $\overline{\overline{\mathbf{N}}} \leq \overline{\overline{\bigcup_{A \in \mathscr{A}} A}}$.

For each $a \in \bigcup_{A \in \mathscr{A}} A$, there is a unique n such that $a \in A_n$ and a unique $k \in \mathbf{N}$ such that $f_n(k) = a$. Thus we may define $g : \bigcup_{A \in \mathscr{A}} A \to \mathbf{N}$ by setting $g(a) = g(f_n(k)) = 2^n 3^k$. Since the prime factorization of natural numbers is unique, g is one-to-one. Hence $\overline{\overline{\bigcup_{A \in \mathscr{A}} A}} \leq \overline{\overline{\mathbf{N}}}$. By the Cantor-Schröder-Bernstein Theorem, $\bigcup_{A \in \mathscr{A}} A$ is denumerable.

Finally, in the case that the A_i's are not pairwise disjoint, we may consider the denumerable collection of disjoint sets $\{A_i \times \{i\} : i \in \mathbf{N}\}$. By the first part of this proof, $\bigcup_{i \in \mathbf{N}} (A_i \times \{i\})$ is denumerable. Define $h : \bigcup_{i \in \mathbf{N}} (A_i \times \{i\}) \to \bigcup_{i \in \mathbf{N}} A_i$ by letting the image of (a, i) be a. We leave the proof that h is onto $\bigcup_{i \in \mathbf{N}} A_i$ to exercise 4. Thus, by Theorem 5.22, $\overline{\overline{\mathbf{N}}} = \overline{\overline{A_1}} \leq \overline{\overline{\bigcup_{i \in \mathbf{N}} A_i}} \leq \overline{\overline{\bigcup_{i \in \mathbf{N}} (A_i \times \{i\})}} = \overline{\overline{\mathbf{N}}}$, so that $\bigcup_{i \in \mathbf{N}} A_i$ is also denumerable. ■

Example. For each $n \in \mathbf{N}$, let $A_n = \{a/n : a \in \mathbf{N}\}$. Then $\bigcup_{n \in \mathbf{N}} A_n = \mathbf{Q}^+$. Since each A_n is denumerable, Theorem 5.28 provides an alternate proof that $\mathbf{Q}^+$ is denumerable (Theorem 5.12).

Many of the results of this chapter can be amalgamated into our final theorem on cardinality. In terms of the Infinite Hotel it will tell us that even though the countable number of rooms in the hotel each has one person to a room, it is possible to accommodate an additional countable number of people each night for the next countable number of nights.

Theorem 5.29. The union of a countable collection of countable sets is countable.

Proof. The only cases that remain unproved are the denumerable union of a collection of finite sets and the denumerable union of a collection of countable sets (exercise 5). ■

Exercises 5.5

1. Give an example, if possible, of
   ★ (a) a denumerable collection of finite sets whose union is denumerable.
   (b) a denumerable collection of finite sets whose union is finite.
   (c) a denumerable collection of pairwise disjoint finite sets whose union is finite.
2. Prove Corollary 5.25.
3. Prove Theorem 5.27.
4. Complete the proof of Theorem 5.28 by showing that the function h maps onto $\bigcup_{i \in \mathbf{N}} A_i$.
5. Complete the proof of Theorem 5.29.
6. ☆ (a) Prove $\mathbf{N} \times \mathbf{N}$ is denumerable.
   (b) If A and B are denumerable, prove $A \times B$ is denumerable.
   (c) If A and B are countable, prove $A \times B$ is countable.
7. ☆ Use exercise 6 and the Cantor-Schröder-Bernstein Theorem to give an alternate proof that $\mathbf{Q}^+$ is denumerable.
8. Prove:
   (a) the set of all singleton subsets of $\mathbf{N}$ is denumerable.
   ☆ (b) the set of all two-element subsets of $\mathbf{N}$ is denumerable.
   (c) the set of all n-element subsets of $\mathbf{N}$ is denumerable.
   (d) the set of all finite subsets of $\mathbf{N}$ is denumerable.
9. ☆ Let S' be the set of all sequences of zeros and ones where all but a finite number of terms are 0. (See exercise 10 of section 5.2.) Prove S' is denumerable.
10. Prove that if B is uncountable and $\overline{\overline{B}} \le \overline{\overline{A}}$, then A is uncountable.
11. **Proofs to Grade.**
    ★ (a) **Claim.** If A is denumerable then $A - \{x\}$ is denumerable.
    "Proof." Assume A is denumerable.
    Case 1. If $x \notin A$, then $A - \{x\} = A$, which is denumerable by hypothesis.
    Case 2. Assume $x \in A$. Since A is denumerable, there exists $f\colon \mathbf{N} \xrightarrow[\text{onto}]{1-1} A$. Define g by setting $g(n) = f(n+1)$. Then $g\colon \mathbf{N} \xrightarrow[\text{onto}]{1-1} A - \{x\}$, so $\mathbf{N} \approx A - \{x\}$. Therefore $A - \{x\}$ is denumerable. ■
    (b) **Claim.** If A and B are denumerable, then $A \times B$ is denumerable.
    "Proof." Assume A and B are denumerable, but that $A \times B$ is not denumerable. Then $A \times B$ is finite. Since A and B are denumerable, they are not empty; therefore, choose $a \in A$ and $b \in B$. By exercise 3 of section 5.1, $A \approx A \times \{b\}$, and by an obvious modification of that exercise, $B \approx \{a\} \times B$. Since $A \times B$ is finite, the subsets $A \times \{b\}$ and $\{a\} \times B$ are finite. Therefore, A and B are finite. This contradicts the statement that A and B are denumerable. We conclude that $A \times B$ is denumerable. ■

Concepts of Algebra 6

Much of modern mathematics is algebraic in nature. We have in mind a broad meaning of algebra as a system of computation and the study of properties of such a system. In this chapter we make precise the idea of an algebraic structure; study an especially important structure, the group; and investigate the notions of substructure and quotient structure.

SECTION 6.1. ALGEBRAIC STRUCTURES

Let A be a nonempty set. A **binary operation on A** is a function from $A \times A$ to A. We will usually denote an operation by one of the symbols $+$, $\cdot$, $\circ$, or $*$. If the operation is $\circ$ and the image $\circ\,(x, y)$ of the ordered pair (x, y) is z, we usually write $x \circ y = z$. This notation is familiar from the operations of addition and multiplication on the set of real numbers, where it is more natural to write $4 + 7 = 11$ than $+(4, 7) = 11$. Often we omit completely the operation symbol and write $xy = z$, as is done with multiplication.

The images xy, $x \circ y$, and $x * y$ are usually called products, regardless of whether the operations involved have anything to do with multiplication. Similarly, $s + t$ is referred to as the sum of s and t, even when the function $+$ has nothing to do with addition.

Besides the usual arithmetic operations on sets of numbers, you are already familiar with several binary operations. Composition is an operation on the set of all functions that map a set A onto itself. The operation $\cup$, union, is a binary operation on the power set of A. Intersection, set difference, and symmetric difference are other operations on $\mathscr{P}(A)$.

There are operations more complicated than binary operations. Ternary operations, for example, map $A \times A \times A$ to A. An **algebraic structure** or **system** is a nonempty set A together with a collection of (at least one) operations on A and a collection (possibly empty) of relations on A. For example, the system of real numbers with addition and multiplication and the relation "less than" is an algebraic structure. We could as well consider the system of reals with addition, reals with multiplication, or reals with addition, "less than," and "greater than." In this chapter we will restrict ourselves to structures with one binary operation.

Definition. The set A is said to **be closed under** $*$ iff for all $x, y \in A$, $x * y \in A$. The statements "A is closed under $*$," "$*$ is an operation on A," and "$(A, *)$ is an algebraic system" are all equivalent.

Definition. If A is a finite set, the **order** of $(A, *)$ is $\overline{\overline{A}}$, which is the number of elements in A. When A is infinite, we simply say $(A, *)$ has infinite order.

A convenient way to display information about an operation, at least for a system of small finite order, is by means of its **operation table,** or **Cayley table.** An operation table for a system $(A, *)$ of order n is an $n \times n$ array of products such that $x * y$ appears in row x and column y. Table 6.1 represents a system $(A, *)$ with $A = \{1, 2, 3\}$ in which, for example, $2 * 1 = 3$. As an example of computation on this system, notice that $(3 * 2) * (1 * 3) = 3 * 1 = 2$.

TABLE 6.1

$*$	1	2	3
1	3	2	1
2	3	1	3
3	2	3	3

Definitions. Let $(A, *)$ be an algebraic system. Then

(i) $*$ is **commutative** on A iff for all $x, y \in A$, $x * y = y * x$.
(ii) $*$ is **associative** on A iff for all $x, y, z \in A$, $(x * y) * z = x * (y * z)$.
(iii) an element e of A is an **identity element** for $*$ iff for all $x \in A$, $x * e = e * x = x$.
(iv) if A has an identity element e, and a and b are in A, then b is an **inverse** of a iff $a * b = b * a = e$. In this case, a would also be an inverse of b.

You are familiar with the fact that the system $(\mathbf{Z}, \cdot)$ with the usual multiplication of integers is commutative and associative with identity element 1. In this system, only the elements 1 and -1 have inverses. For the system consisting of the real numbers with addition, the operation is commutative and associative, 0 is the identity, and every element has an inverse (its negative).

Our study of functions provides the information needed to consider what can be regarded as the most important kind of algebraic system. The set of all functions mapping a set one-to-one onto itself is closed under the operation of function composition. This operation is associative, but generally not commutative. The identity function is the identity element, and every element f has an inverse, f^{-1} (see Theorem 4.9).

The operation $*$ of table 6.1 is not commutative because, for example, $1 * 3 \neq 3 * 1$. It is not associative because $(1 * 1) * 2 \neq 1 * (1 * 2)$. There is no identity element, so the question of inverses does not even arise.

Three different operations on $A = \{1, 2, 3\}$ are shown in tables 6.2, 6.3, and 6.4.

TABLE 6.2

$\circ$	1	2	3
1	1	2	3
2	1	2	3
3	1	2	3

TABLE 6.3

$\cdot$	1	2	3
1	3	1	2
2	1	2	3
3	2	3	1

TABLE 6.4

+	1	2	3
1	3	3	1
2	1	1	2
3	1	2	3

Tables 6.2 and 6.4 are not commutative. The fact that the operation of table 6.3 is commutative on A can be seen by noting that the table is symmetric about its **main diagonal,** from the upper left to lower right.

It is not easy to tell by looking at a table whether an operation is associative. For a system of order n, verification of associativity may require checking n^3 products of three elements, each grouped two ways. The operations in tables 6.2 and 6.3 are associative, but $+$ is not associative on A. (Why?)

The associative property is a great convenience in computing products. First, it means that as long as factors appear in the same order, we need no parentheses. For both $x(yz)$ and $(xy)z$ we can write xyz. This can be extended inductively to products of four or more factors: $(xy)(zw) = (x(yz)w) = x(y(zw))$, and so forth. Second, for an associative operation we can define powers. Without associativity, $(xx)x$ might be different from $x(xx)$, but with associativity they are equal, and both can be denoted by x^3.

The element 2 is an identity for $\cdot$ of table 6.3, and in this system the inverses of 1, 2, and 3 are, respectively, 3, 2, and 1. Table 6.2 has no identity element. In table 6.4, where 3 is the identity element, only 1 and 3 have inverses.

One of the most important concepts in algebra involves mappings between systems. Most interesting are the **operation preserving mappings.** As we shall see, such mappings actually preserve the structure of an algebraic system.

> **Definition.** Let $(A, \circ)$ and $(B, *)$ be algebraic systems. The mapping $f: A \to B$ is called **operation preserving** (or OP) iff for all $x, y \in A$, $f(x \circ y) = f(x) * f(y)$.

A simplified statement of how an OP map works is "the image of a product is the product of the images." That is, the result is the same whether operating or mapping is done first. That "the image of the product is the product of the images" must be carefully understood. To compute $f(x \circ y)$, one must use the operation of $(A, \circ)$ while in $f(x) * f(y)$, the operation is the operation of $(B, *)$. The final equation is an equality of elements of B, because $f(x \circ y)$ and $f(x) * f(y)$ are elements of B.

Examples of OP maps abound and mathematicians delight in finding them. The familiar equation

$$\log (x \cdot y) = \log x + \log y$$

tells us that the logarithm function from $(\mathbf{R}^+, \cdot)$ to $(\mathbf{R}, +)$ is operation preserving. As another example, let $(P, +)$ be the set of polynomial functions on $\mathbf{R}$ with the operation of function addition. Then the differentiation map $D: (P, +) \rightarrow (P, +)$ defined by $D(F) = F'$, the first derivative of F, is operation preserving. This mapping preserves addition because $D(F + G) = (F + G)' = F' + G' = D(F) + D(G)$. In other words, the derivative of the sum is the sum of the derivatives. This same mapping from $(P, \cdot)$ onto itself, where $\cdot$ is the operation of function multiplication, is not operation preserving, because $D(F \cdot G)$ is not usually equal to $D(F) \cdot D(G)$.

The next series of theorems is a part of the explanation of what we mean by saying that an OP map preserves the structure of an algebra. We will see that under an OP map the image set (range) of an algebraic system is an algebraic system and that, if the original operation is commutative or has an identity element, then so does the image set. More information about OP maps appears in later sections.

Theorem 6.1. Let f be an OP map from $(A, \circ)$ to $(B, *)$. Then $(\text{Rng}(f), *)$ is an algebraic structure.

Proof. ⟨*What we must show is that $*$ is closed on $\text{Rng}(f)$; that is, if $u, v \in \text{Rng}(f)$, then $u * v \in \text{Rng}(f)$.*⟩

Assume $u, v \in \text{Rng}(f)$. Then there exist elements x and y of A such that $f(x) = u$ and $f(y) = v$. Then $u * v = f(x) * f(y) = f(x \circ y)$, so $u * v$ is the image of $x \circ y$, which is in A. Therefore, $u * v \in \text{Rng}(f)$. ■

Now that we know that the range of an OP map is an algebraic system, we can simplify things by ignoring the part of the codomain that is not in the range. This amounts to assuming that f maps onto B. It does not mean that every OP map is onto every possible codomain.

Theorem 6.2. Let f be an OP map from $(A, \circ)$ onto $(B, *)$. If $\circ$ is commutative on A, then $*$ is commutative on B.

Proof. Assume that f is OP and $\circ$ is commutative on A. Let u and v be elements of B. ⟨*We want to show $u * v = v * u$.*⟩

From the fact that f maps onto B we know that there are x and y in A such that $u = f(x)$ and $v = f(y)$. Then $u * v = f(x) * f(y) =$

$f(x \circ y) = f(y \circ x) = f(y) * f(x) = v * u$. ⟨*This equation uses the fact that* $\circ$ *is commutative and (twice) that* f *is OP.*⟩ Therefore, $u * v = v * u$. ■

The properties of associativity, existence of identities, and existence of inverses are all preserved by an OP mapping.

Theorem 6.3. Let f be an OP map from $(A, \circ)$ onto $(B, *)$.

(a) If $\circ$ is associative on A, then $*$ is associative on B.
(b) If e is an identity for A, then $f(e)$ is an identity for B.
(c) If x^{-1} is an inverse for x in A, then $f(x^{-1})$ is an inverse for $f(x)$ in B.

Proof. Exercise 7. ■

Exercises 6.1

1. Which of the following are algebraic structures? (The symbols $+$, $-$, $\cup$, $\cap$, and so on, have their usual meanings.)
★ (a) $(\mathbf{Z}, -)$ (b) $(\mathbf{Z}, \div)$ (c) $(\mathbf{R}, -)$
(d) $(\mathbf{R}, \div)$ ★ (e) $(\mathbf{N}, -)$ (f) $(\mathbf{Q}, \div)$
(g) $(\mathbf{Q} - \{0\}, \div)$ (h) $(\mathscr{P}(A), \cap)$ (i) $(\mathscr{P}(A), \cup)$
(j) $(\mathscr{P}(A) - \{\varnothing\}, -)$
☆ 2. Which of the operations in exercise 1 are commutative?
☆ 3. Which of the operations in exercise 1 are associative?
4. Let $\mathscr{M} = \{A : A \text{ is an } m \times n \text{ matrix with real entries}\}$.
(a) If $\cdot$ is matrix multiplication, is $(\mathscr{M}, \cdot)$ an algebraic system?
(b) If $+$ is matrix addition, is $(\mathscr{M}, +)$ an algebraic system?
5. The Cayley tables for operations $\circ$, $*$, $+$, $\times$ are listed below.

$\circ$	a	b	c	d
a	a	b	c	d
b	b	a	d	c
c	c	d	a	b
d	d	c	b	a

$*$	a	b	c
a	c	a	c
b	a	b	c
c	b	c	b

$+$	a	b
a	a	a
b	a	a

$\times$	a	b	c
a	a	c	b
b	c	b	a
c	b	a	c

(a) Which of the operations are commutative?
(b) Which of the operations are associative?
(c) Which systems have an identity? What is the identity element?
(d) For those systems that have an identity, which elements have inverses?
6. Prove that if $(A, *)$ is an algebraic system and $*$ is associative on A, then the product of any n elements from A can be grouped in any manner we wish (as long as the order of the factors is unchanged) without affecting the product. *Hint:* Use complete induction to prove that the product of $a_1, a_2, a_3, \ldots, a_n$ is equal to $(\cdots((a_1 * a_2) * a_3)\cdots) * a_n$.
7. Prove Theorem 6.3.
☆ 8. Let $(A, \circ)$ be an algebraic structure. Prove that if e and f are both identities for $\circ$, then $e = f$.
9. Let $(A, \circ)$ be an algebraic structure, $a \in A$, and e the identity for $\circ$.
☆ (a) Prove that if $\circ$ is associative, and x and y are inverses of a, then $x = y$.
(b) Give an example of a nonassociative structure in which inverses are not unique.

10. Let $(A, \circ)$ be an algebraic structure. An element $\ell \in A$ is a *left identity* for $\circ$ iff $\ell \circ a = a$ for every $a \in A$.
    (a) Give an example of a structure of order 3 with exactly two left identities.
    (b) Define a right identity for $(A, \circ)$.
    (c) Prove that if $(A, \circ)$ has a right identity r and left identity ℓ, then $r = \ell$, and that $r = \ell$ is an identity for $\circ$.
11. Define SQRT: $(\mathbf{R}^+, +) \to (\mathbf{R}^+, +)$ by SQRT $(x) = \sqrt{x}$. Is SQRT operation preserving?
12. Is SQRT: $(\mathbf{R}^+, \cdot) \to (\mathbf{R}^+, \cdot)$ operation preserving?
13. Is SQR: $(\mathbf{R}, +) \to (\mathbf{R}, +)$ defined by SQR $(x) = x^2$ operation preserving?
14. Is SQR: $(\mathbf{R}, \cdot) \to (\mathbf{R}, \cdot)$ operation preserving?

★ 15. Let $\mathscr{F}$ be the set of all real valued integrable functions defined on the interval $[a, b]$. Then $(\mathscr{F}, +)$ is an algebraic structure, where $+$ is ordinary addition of functions. Define $I: (\mathscr{F}, +) \to (\mathbf{R}, +)$ by $I(f) = \int_a^b f(x)dx$. Use your knowledge of calculus to verify that I is an OP map.
16. Let $\mathscr{F}$ be as in exercise 15. Consider the algebraic structures $(\mathscr{F}, \cdot)$ and $(\mathbf{R}, \cdot)$ where $\cdot$ represents ordinary function product and real number multiplication, respectively. Verify that $I: (\mathscr{F}, \cdot) \to (\mathbf{R}, \cdot)$, defined by $I(f) = \int_a^b f(x)dx$, is not operation preserving.
17. Let $f: (A, \circ) \to (B, *)$ and $g: (B, *) \to (C, \times)$ be OP maps.
    (a) Prove $g \circ f$ is an OP map.
    (b) Prove that if f is a one-to-one correspondence, then f^{-1} is an OP map.
18. Let $\mathscr{M}$ be the set of all 2×2 matrices with real entries. Define Det : $\mathscr{M} \to \mathbf{R}$ by

$$\text{Det} \begin{bmatrix} a & b \\ c & d \end{bmatrix} = ad - bc.$$

    (a) Let $(\mathscr{M}, \cdot)$ represent $\mathscr{M}$ with matrix multiplication and $(\mathbf{R}, \cdot)$ represent the reals with ordinary multiplication. Prove Det : $(\mathscr{M}, \cdot) \to (\mathbf{R}, \cdot)$ is operation preserving.
    (b) Let $(\mathscr{M}, +)$ represent $\mathscr{M}$ with matrix addition. Prove Det : $(\mathscr{M}, +) \to (\mathbf{R}, +)$ is not operation preserving.
19. (a) Let $\mathbf{C}$ denote the complex numbers with ordinary addition. Define Conj : $(\mathbf{C}, +) \to (\mathbf{C}, +)$ by Conj $(a + bi) = a - bi$. Verify that Conj is an OP map.
    (b) Verify that Conj : $(\mathbf{C}, \cdot) \to (\mathbf{C}, \cdot)$ is an OP map.
20. Let $t \in \mathbf{R}$. Define $S_t: (\mathbf{R}, +) \to (\mathbf{R}, +)$ by $S_t(x) = tx$. Prove that S_t is operation preserving.
21. Let $f: A \to B$.
    ★ (a) Verify that the induced function $f: (\mathscr{P}(A), \cup) \to (\mathscr{P}(B), \cup)$ is an OP map.
    (b) Verify that the induced function $f^{-1}: (\mathscr{P}(B), \cap) \to (\mathscr{P}(A), \cap)$ is an OP map.
    (c) Verify that the induced function $f^{-1}: (\mathscr{P}(B), \cup) \to (\mathscr{P}(A), \cup)$ is an OP map.
22. (a) **Claim.** If $(A, \circ)$ is an algebraic structure, if e is an identity for $\circ$, and if x and y are both inverses of a, then $x = y$.
    "Proof." $x = x \circ e = x \circ (a \circ y) = (x \circ a) \circ y = e \circ y = y$. Thus $x = y$ and inverses are unique. ■
    ★ (b) **Claim.** If every element of a structure $(A, \circ)$ has an inverse, then $\circ$ is commutative.
    "Proof." Let x and y be in A. The element y has an inverse, which we will call y'. Then $y \circ y' = e$, so y is the inverse of y'. Now $x = x$, and multiplying both sides of this equation by the inverse of y', we have $y \circ x = x \circ y$. Therefore $\circ$ is commutative. ■

SECTION 6.2. GROUPS

We have considered algebraic structures $(A, \circ)$ with one binary operation. For the rest of this chapter we will give our attention to certain of these structures, the groups. The properties of commutativity, associativity, and existence of identities and inverses examined in section 6.1 are just the properties we consider in defining a group. Our approach is **axiomatic.** That is, we shall list the desired properties (axioms) of a structure, and any system having these properties is called a group. There are some important observations to be made about such a method. Although stated for groups, the following comments apply generally to axiomatic studies.

First, a small set of axioms is advantageous (and may be challenging to produce) because a small set means that few properties need be checked to be sure a structure is a group. It may be best to leave a desired property out of the axioms if it can be deduced from the remaining axioms. This is so because every consequence of the axioms must be true of all structures satisfying the axioms. Finally, the fact that axiom systems may be altered by adding or deleting specific axioms does not mean that axioms are chosen at random, or that all axioms are equally worthy of study. The group axioms are chosen because the structures they describe are central to modern mathematics.

It was the work of Evariste Galois (1811–1832) on solving polynomial equations that led to the study of groups. The concept of a group has influenced and enriched other areas of mathematics. In geometry, for example, the ideas of Euclidean and non-Euclidean geometries were unified by the notion of the group. Group theory has been applied outside mathematics, too, in fields such as nuclear physics and crystallography.

Definition. A **group** is an algebraic structure $(G, \circ)$ such that:

(1) the operation $\circ$ is associative,
(2) there is an identity element $e \in G$ for $\circ$,
(3) every $x \in G$ has an inverse x^{-1} in G.

To prove that a structure is a group we prove that axioms 1, 2, and 3 hold for the structure. Implicitly, we must also verify that $(G, \circ)$ is actually an algebraic structure. That is, we must know that $\circ$ satisfies the closure property: if $x, y \in G$, then $x \circ y \in G$.

A fourth axiom may be considered. A group is called **abelian** iff

(4) the operation $\circ$ is **commutative.**

The fourth axiom is an **independent** axiom because it cannot be proved from the other group axioms.

The systems $(\mathbb{R}, +)$, $(\mathbb{Q}, +)$, and $(\mathbb{Z}, +)$ are all groups. The system $(\mathbb{R}, \cdot)$ is not a group because 0 has no inverse, but $(\mathbb{R} - \{0\}, \cdot)$ and $(\mathbb{Q} - \{0\}, \cdot)$ are groups. The systems $(\mathbb{N}, +)$, $(\mathbb{Z}, -)$, and $(\mathbb{Z} - \{0\}, \cdot)$ are not groups

(Why?), but $(\{0\}, +)$ and $(\{1, -1\}, \cdot)$ are groups. Table 6.5 is the Cayley table for $(\{1, -1\}, \cdot)$.

TABLE 6.5

$\cdot$	1	-1
1	1	-1
-1	-1	1

The group axioms do not require that the identity and the inverses of elements be unique. However, that the identity and the inverses are unique can be proved. (See exercises 8 and 9 in section 6.1.) Leaving the uniqueness condition out of the group axioms shortens the verification that a structure is a group.

Frequently we will simply say "the group G" meaning the set G where the product of elements x and y is denoted xy. Two elementary consequences of the group axioms facilitate calculations involving elements of a group.

Theorem 6.4. If G is a group and $a, b \in G$, then $(a^{-1})^{-1} = a$ and $(ab)^{-1} = b^{-1}a^{-1}$.

Proof. Because a^{-1} is the inverse of a, $a^{-1}a = aa^{-1} = e$. Therefore a acts as the ⟨*unique*⟩ inverse of a^{-1}, so $(a^{-1})^{-1} = a$.

We know $(ab)^{-1}$ is the unique element x of G such that $(ab)x = x(ab) = e$. We see that $b^{-1}a^{-1}$ meets this criterion by computing $(ab)(b^{-1}a^{-1}) = a(bb^{-1})a^{-1} = a(e)a^{-1} = aa^{-1} = e$. Similarly, $(b^{-1}a^{-1})(ab) = e$, so $b^{-1}a^{-1}$ is the inverse of ab. ■

Theorem 6.5. Let G be a group. Then left and right cancellation both hold in G. That is, for elements x, y, z of G,

(a) if $xy = xz$, then $y = z$,
(b) if $yx = zx$, then $y = z$.

Proof.

(a) Suppose $xy = xz$ in the group G. Then, using the fact that $x^{-1} \in G$, we have $x^{-1}(xy) = x^{-1}(xz)$. Therefore, using the associative, inverse, and identity properties, we see that $x^{-1}(xy) = (x^{-1}x)y = ey = y = x^{-1}(xz) = (x^{-1}x)z = ez = z$. Therefore $y = z$. This proves (a), and the proof of (b) is similar. (Exercise 6.) ■

It follows from (a) of Theorem 6.5 that no element can occur twice in any row of the operation table for a group. In fact if a and b are in G, then $a(a^{-1}b) = b$, so every element b occurs in row a, for every a. Thus, if G is a group, then every element of G occurs exactly once in every row and (using (b)) in every column of the operation table. The converse of this statement is false.

If G is a group, it is convenient to have notation for powers of elements of G. Let $a \in G$ and $n \in \mathbf{N}$. We define inductively $a^0 = e$ and $a^{n+1} = a^n a$. Thus, a^n is defined inductively for all $n \geq 0$. Define a^n for $n < 0$ by $a^n = (a^{-1})^{-n}$. Then we can prove that for $n > 0$, $(a^n)^{-1} = a^{-n}$. Now we have the usual laws of exponents for $m, n \in \mathbf{N}$:

$$a^n a^m = a^{n+m}$$
$$(a^n)^m = a^{nm}$$

If and only if the group is abelian, we also have

$$(ab)^n = a^n b^n.$$

When the group operation is addition, we usually write the inverse of an element a as $-a$ and call it the **negative** of a. Rather than powers of a, we refer to **multiples** of a, but the only difference between these ideas is notation. Thus $0a = e$ and $(n + 1)a = na + a$ for $n > 0$ and $na = -n(-a)$ for $n < 0$.

Definition. An operation preserving map from the group $(G, \circ)$ to the group $(H, *)$ is called a **homomorphism** from $(G, \circ)$ to $(H, *)$.

A homomorphism is just an OP map for groups. A **homomorphic image** is the range of a homomorphism. The function $f: (\mathbf{Z}, +) \rightarrow (\mathbf{Z}, +)$ given by $f(x) = 3x$ is an example of a homomorphism. The corresponding homomorphic image is $\text{Rng}(f) = \{3x: x \in \mathbf{Z}\} = \{\ldots, -6, -3, 0, 3, 6, 9, \ldots\}$, which is itself a group under the operation of addition.

Theorem 6.6.

(a) The homomorphic image of a group is a group.
(b) The homomorphic image of an abelian group is an abelian group.

Proof. The proof is nothing more than an observation that the image of a group under a homomorphism is an algebraic system ⟨*Theorem 6.1*⟩ that is associative, has an identity element, and has an inverse for every element ⟨*Theorem 6.3*⟩. Furthermore, if $(G, \circ)$ is abelian, then the image is also abelian ⟨*Theorem 6.2*⟩. ■

Let $(G, \circ)$ be any group with identity e. Define a mapping $F: G \rightarrow G$ by setting $F(x) = e$, for every $x \in G$. Then F is a homomorphism on G. We can verify this by observing that for x and $y \in G$, $F(x \circ y) = e$ and $F(x) \circ F(y) = e \circ e = e$, so $F(x \circ y) = F(x) \circ F(y)$. In this case the group that is the image of the homomorphism is $(\{e\}, \circ)$.

Exercises 6.2

1. Show that each of the following algebraic structures is a group.
   ★ (a) $(\{1, -1, i, -i\}, \cdot)$ where $i = \sqrt{-1}$ and $\cdot$ is complex number multiplication.
   (b) $(\{1, \alpha, \beta\}, \cdot)$ where $\alpha = \dfrac{-1 + i\sqrt{3}}{2}$, $\beta = \dfrac{-1 - i\sqrt{3}}{2}$, and $\cdot$ is complex number multiplication.
   (c) $(\{1, -1\}, \cdot)$ where $\cdot$ is multiplication of integers.
   ☆ (d) $(\mathscr{P}(X), \Delta)$ where X is a nonempty set and $A \,\Delta\, B = (A - B) \cup (B - A)$.
2. ★ Given that $G = \{e, u, v, w\}$ is a group of order 4 with identity e and $u^2 = v$, $v^2 = e$, construct, if possible, the operation table for G.
3. Given that $G = \{e, u, v, w\}$ is a group of order 4 with identity e and $e^2 = u^2 = v^2 = w^2 = e$, construct, if possible, the operation table for G.
4. ☆ Which of the groups of exercise 1 are abelian?
5. By Theorem 6.4, $(ab)^{-1} = b^{-1}a^{-1}$.
   (a) Prove that $(abc)^{-1} = c^{-1}b^{-1}a^{-1}$.
   (b) Prove a similar result for $(a_1a_2a_3 \dots a_n)^{-1}$ by induction.
6. Prove part (b) of Theorem 6.5. That is, prove that if G is a group, x, y, and z are elements of G, and if $yx = zx$, then $y = z$.
7. ☆ Let G be a group. Prove that if $g^2 = e$ for all $g \in G$, then G is abelian.
8. ☆ Give an example of an algebraic structure of order 4 that has both right and left cancellation but that is not a group.
9. Let G be an abelian group and let $a, b \in G$.
   (a) Prove that $a^2b^2 = (ab)^2$.
   (b) Prove that for all $n \in \mathbf{N}$, $a^nb^n = (ab)^n$.
10. Show that the structure $(\mathbf{R} - \{1\}, \circ)$, with operation $\circ$ defined by $a \circ b = a + b - ab$, is an abelian group. (You should first show that $(\mathbf{R} - \{1\}, \circ)$ is indeed a structure.)
11. ☆ Let $(G, *)$ be a group and $a, b \in G$. Show that there exist unique elements x and y in G such that $a * x = b$ and $y * a = b$.
12. Show that $(\mathbf{Z}, \#)$ with operation # defined by $a \# b = a + b + 1$ is a group. Find x such that $50 \# x = 100$.
13. **Proofs to Grade.**
   (a) **Claim.** If G is a group, then G is commutative.
   "Proof." Let a and b be elements of G. Then

$$\begin{aligned}
ab &= a\,e\,b \\
&= a(ab)(ab)^{-1}b \\
&= a(ab)(b^{-1}a^{-1})b \\
&= (aa)(bb^{-1})a^{-1}b \\
&= (aa)a^{-1}b \\
&= (aa)(b^{-1}a)^{-1} \\
&= a(ab^{-1})a^{-1} \\
&= a(a^{-1}b)a \\
&= (aa^{-1})(ba) \\
&= e\,(ba) \\
&= ba.
\end{aligned}$$

   Therefore, $ab = ba$, and G is commutative. ■

   ★ (b) **Claim.** If G is a group with elements x, y, and z, and if $xz = yz$, then $x = y$.
   "Proof." If $z = e$, then $xz = yz$ implies that $xe = ye$, so $x = y$. If $z \neq e$, then the inverse of z exists, and $xz = yz$ implies $xz/z = yz/z$ and $x = y$. Hence, in all cases, if $xz = yz$, then $x = y$. ■

SECTION 6.3. EXAMPLES OF GROUPS

A function on a set A which is one-to-one and onto A is called a **permutation of A.** We saw in section 6.1 that the structure consisting of the set of all permutations of A with the operation of composition has the properties of a group. The group we have described is called the **group of permutations of A.** Groups whose elements are some (but not necessarily all) permutations of a set are called **permutation groups.** The reason for the importance of permutation groups is that, for every group with elements of any kind (numbers, sets, functions), there is a corresponding group of permutations with the same structure as the original group. (See Theorem 6.22.)

Definition. For $n \in \mathbf{N}$, the group of all permutations of a set with n elements is called the **symmetric group on n symbols** and is designated by S_n.

Let $A = \{1, 2, 3\}$. In order to examine the symmetric group S_3 closely, we shall adopt one of the common notations for a permutation. We think of a function f on A as follows:

$$\begin{matrix} \{1 & 2 & 3\} \\ \downarrow & \downarrow & \downarrow \\ \{f(1) & f(2) & f(3)\} \end{matrix}$$

and write f as $(f(1)f(2)f(3))$. Then (123) represents the identity function on A and (213) represents the function g where $g(1) = 2$, $g(2) = 1$, and $g(3) = 3$. We can now list all six elements of S_3 and, after extensive computation, fill in the operation table.

TABLE 6.6: S_3

$\circ$	(123)	(213)	(321)	(132)	(231)	(312)
(123)	(123)	(213)	(321)	(132)	(231)	(312)
(213)	(213)	(123)	(312)	(231)	(132)	(321)
(321)	(321)	(231)	(123)	(312)	(213)	(132)
(132)	(132)	(312)	(231)	(123)	(321)	(213)
(231)	(231)	(321)	(132)	(213)	(312)	(123)
(312)	(312)	(132)	(213)	(321)	(123)	(231)

Recall that the operation $\circ$ here is function composition so that, in order to compute the product of $g = (213)$ and $h = (312)$, we note that $(g \circ h)(1) = g(h(1)) = g(3) = 3$; $(g \circ h)(2) = g(h(2)) = g(1) = 2$; and $(g \circ h)(3) = g(h(3)) = g(2) = 1$. Therefore, $g \circ h = (321)$. This is the first example we have seen of a group that is not abelian.

A symmetry of a geometric figure is a rigid motion (without bending or tearing) of the figure onto itself. More formally, a **symmetry** is a one-to-one distance-preserving transformation of the points of the figure. Every regular polygon and regular solid has an interesting group of symmetries. We consider the symmetries of a square by imagining a cardboard square with vertices 1, 2, 3, 4 and center C. (See figure 6.1.) The square is carried onto itself by the following rigid motions:

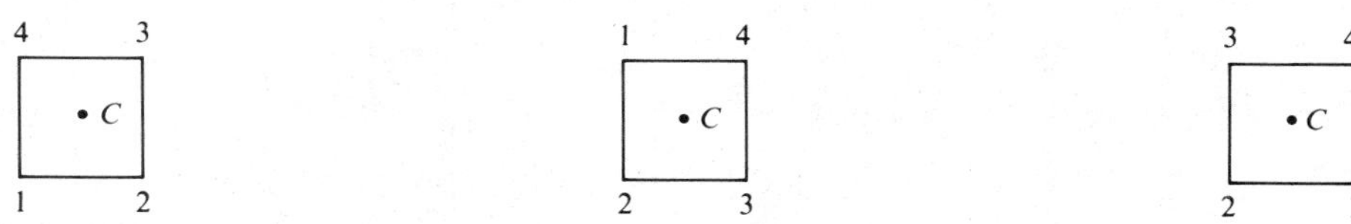

Figure 6.1 **Figure 6.2** **Figure 6.3**

R: a 90° rotation clockwise around center C; R transforms the square to the position shown in figure 6.2.
R^2: a 180° rotation clockwise around C.
R^3: a 270° rotation clockwise around C.
H: a reflection about the horizontal axis through C.
V: a reflection about the vertical axis through C; V transforms the square to the position shown in figure 6.3.
D: a reflection about the lower-left to upper-right diagonal.
D': a reflection about the upper-left to lower-right diagonal.
I: the identity transformation.

We compute the product of two symmetries by performing them in succession. Thus VR is the result of a vertical axis reflection followed by a 90° clockwise rotation. By experimenting with a small cardboard square we find that $VR = D \neq RV = D'$. The table for this permutation group $\mathcal{O}$, called the **octic group,** is given below.

	I	R	R^2	R^3	H	V	D	D'
I	I	R	R^2	R^3	H	V	D	D'
R	R	R^2	R^3	I	D	D'	V	H
R^2	R^2	R^3	I	R	V	H	D'	D
R^3	R^3	I	R	R^2	D'	D	H	V
H	H	D'	V	D	I	R^2	R^3	R
V	V	D	H	D'	R^2	I	R	R^3
D	D	H	D'	V	R	R^3	I	R^2
D'	D'	V	D	H	R^3	R	R^2	I

Recall the notion of congruence modulo m as an equivalence relation on $\mathbf{Z}$. We have seen that for each natural number m there are exactly m equivalence classes, $0/\equiv_m$, $1/\equiv_m, \ldots, (m-1)/\equiv_m$, and the set of all these classes is called $\mathbf{Z}_m$. We define two operations $+_m$ and $\cdot_m$ on the set $\mathbf{Z}_m$ as follows:

$$a/\equiv_m +_m b/\equiv_m = (a+b)/\equiv_m$$

and

$$a/\equiv_m \cdot_m b/\equiv_m = (a \cdot b)/\equiv_m.$$

That is, the sum of the classes is the class of the sum, and so on.

A difficulty must be overcome, however, for we must be sure that $+_m$ and $\cdot_m$ are operations on $\mathbf{Z}_m$. The problem is that $+_m$ and $\cdot_m$ must be functions from $\mathbf{Z}_m \times \mathbf{Z}_m$ to $\mathbf{Z}_m$. Take for example $m = 5$. Then $3/\equiv_5 = -12/\equiv_5$ and $1/\equiv_5 = 31/\equiv_5$, so we must have $3/\equiv_5 +_5 1/\equiv_5 = -12/\equiv_5 +_5 31/\equiv_5$ or else $+_5$ would not have unique images. We must be certain that addition and multiplication as defined in $\mathbf{Z}_m$ do not depend on which elements of the equivalence class are added. This is all taken care of by means of the following.

Lemma 6.7. *If $a \equiv_m b$ and $c \equiv_m d$, then*

(i) $a + c \equiv_m b + d$

and

(ii) $a \cdot c \equiv_m b \cdot d$.

Proof. Exercise 3. ■

In light of the discussion above and Lemma 6.7 we now have that $(\mathbf{Z}_m, +_m)$ and $(\mathbf{Z}_m, \cdot_m)$ are algebraic systems for every $m \in \mathbf{N}$.

Theorem 6.8. *For every natural number m, $(\mathbf{Z}_m, +_m)$ is an abelian group.*

Proof. The proof that $+_m$ is associative is a routine verification using the definition of $+_m$ and the associativity of $+$ for $\mathbf{Z}$. Commutativity is similar. The identity element is $0/\equiv_m$ and the negative of $a/\equiv_m$ is the element $(m-a)/\equiv_m$. ■

When the meaning is clear from the context we will dispense with the cumbersome notation $a/\equiv_m$ for the equivalence class of a modulo m and simply write a. At times we even use just $+$ for $+_m$. The tables for $+_2$, $+_6$, and $\cdot_6$ are shown.

$+_2$	0	1
0	0	1
1	1	0

$+_6$	0	1	2	3	4	5
0	0	1	2	3	4	5
1	1	2	3	4	5	0
2	2	3	4	5	0	1
3	3	4	5	0	1	2
4	4	5	0	1	2	3
5	5	0	1	2	3	4

$\cdot_6$	0	1	2	3	4	5
0	0	0	0	0	0	0
1	0	1	2	3	4	5
2	0	2	4	0	2	4
3	0	3	0	3	0	3
4	0	4	2	0	4	2
5	0	5	4	3	2	1

Notice from its table that $(\mathbf{Z}_6, \cdot_6)$ is not a group and that it is possible to have $a \neq 0$, $b \neq 0$, but $a \cdot_6 b = 0$. In $\mathbf{Z}_m$, if $a \neq 0$ is an element such that $a \cdot b = 0$ for some $b \neq 0$ in $\mathbf{Z}_m$, we say that a is a **divisor of zero.** It can be shown that $(\mathbf{Z}_m - \{0\}, \cdot_m)$ has no divisors of zero iff m is a prime.

Theorem 6.9. The structure $(\mathbf{Z}_m - \{0\}, \cdot_m)$ is an abelian group iff m is prime.

Proof. Exercise 4. ■

Exercises 6.3

1. Construct the operation table for S_2, the symmetric group on 2 elements. Is S_2 abelian?
2. Construct the operation table for S_4, the symmetric group on 4 elements. What is the order of S_4? Is S_4 abelian?

☆ 3. Prove Lemma 6.7.

4. (a) Prove that $(\mathbf{Z}_m - \{0\}, \cdot_m)$ has no divisors of zero iff m is a prime.
   (b) Prove Theorem 6.9.
5. Let $e = (123)$, $\alpha = (231)$, $\beta = (132)$ be permutations in S_3.
   (a) Find α^2.
   (b) Show that $\beta\alpha = \alpha^2\beta$.
   (c) Show that $\{e, \alpha, \alpha^2, \beta, \alpha\beta, \alpha^2\beta\}$ is the set S_3.
   (d) Construct an operation table for S_3 using the symbols in part (c).
6. Let γ be the permutation on $\{1, 2, 3, 4, 5, 6\}$ given by $\gamma = (641253)$. Find
   (a) γ^2. ★ (b) γ^3. (c) γ^6. (d) γ^{-1}.
7. For each of the following permutations α, find $\alpha^{-1}, \alpha^2, \alpha^3, \alpha^4, \alpha^{50}, \alpha^{51}$.
   (a) $\alpha = (21) \in S_2$. ★ (b) $\alpha = (231) \in S_3$. (c) $\alpha = (3412) \in S_4$.
8. List the symmetries of an equilateral triangle and compute four typical products.
9. List the symmetries of a rectangle and give the group table.

★ 10. How many elements are there in the group of symmetries of a regular pentagon? a regular hexagon? a regular n-sided polygon?

11. Construct the operation table for each of the following groups.
    (a) $(\mathbf{Z}_8, +)$. (b) $(\mathbf{Z}_3 - \{0\}, \cdot)$. (c) $(\mathbf{Z}_7 - \{0\}, \cdot)$.
    (d) $(\mathbf{Z}_9 - \{0, 3, 6\}, \cdot)$. (e) $(\mathbf{Z}_{15} - \{0, 3, 5, 6, 9, 10, 12\}, \cdot)$.
12. Find all zero divisors in the algebraic structure
    ★ (a) $(\mathbf{Z}_{12}, \cdot)$ (b) $(\mathbf{Z}_{15}, \cdot)$ (c) $(\mathbf{Z}_m, \cdot)$

★ 13. If p is prime, show that $(p - 1)^{-1} = p - 1$ in $(\mathbf{Z}_p - \{0\}, \cdot)$.

14. Find all solutions in $(\mathbf{Z}_{20}, \cdot)$ for the following equations:
    ★ (a) $5 \cdot x = 0$. (b) $3 \cdot x = 0$.
    (c) $x \cdot x = 0$. (d) $x^2 - 6x + 5 = 0$.
15. **Proofs to Grade.**
    (a) **Claim.** If a and b are zero divisors in $(\mathbf{Z}_m, \cdot)$, then ab is a zero divisor.
    "Proof." If a and b are zero divisors, then $ab = 0$. Thus $(ab)(ab) = 0 \cdot 0 = 0$ and ab is a zero divisor. ■
    ★ (b) **Claim.** If a and b are zero divisors in $(\mathbf{Z}_m, \cdot)$ and $ab \neq 0$, then ab is a zero divisor.
    "Proof." Since a is a zero divisor, $ax = 0$ for some $x \neq 0$ in $\mathbf{Z}_m$. Likewise $by = 0$ for some $y \neq 0$ in $\mathbf{Z}_m$. Therefore $(ab)(xy) = (ax)(by) = 0 \cdot 0 = 0$. Thus ab is a zero divisor. ■

SECTION 6.4. SUBGROUPS

In general, a substructure of an algebraic system $(A, *)$ consists of a subset of A together with all the operations and relations in the original structure, *provided* that this is an algebraic structure. This proviso is necessary, for it may happen that a subset of A is not closed under an operation. For example, the subset of $\mathbf{R}$ consisting of the irrationals is not closed under multiplication. Substructures are a natural idea, and they can be useful in describing a structure.

Definition. Let $(G, \circ)$ be a group and H a subset of G. Then $(H, \circ)$ is a **subgroup** of G iff $(H, \circ)$ is a group.

It is understood that the operation $\circ$ on H agrees with the operation $\circ$ on G. That is, the operation on H is the function $\circ$ restricted to $H \times H$.

If $H \subseteq G$ and $(G, \circ)$ is a group, then in order to prove H is a subgroup of G all three of the group properties and closure must be proved for $(H, \circ)$. At least, that will be the situation until we find some way to shorten the work.

For each integer m, the set of all multiples of m is a subgroup of $(\mathbf{Z}, +)$. $(\mathbf{Z}, +)$ is a subgroup of $(\mathbf{Q}, +)$. Both $\{0, 3\}$ and $\{0, 2, 4\}$ are subgroups of $(\mathbf{Z}_6, +)$. The tables for these two groups are shown.

+	0	3
0	0	3
3	3	0

+	0	2	4
0	0	2	4
2	2	4	0
4	4	0	2

For every group $(G, \circ)$ with identity e, $(\{e\}, \circ)$ is a group. This group is called the **identity subgroup,** or **trivial subgroup** of G. Also, every group is a subgroup of itself. All subgroups of G other than G and $\{e\}$ are called **proper subgroups.**

Important questions to be answered are whether the identity element in a subgroup can be different from the identity of the original group, and whether the inverse of an element in H could be different from its inverse in G. The answers are "no" and "no."

Theorem 6.10. Let H be a subgroup of G. The identity of H is the identity e of G and if $x \in H$, the inverse of x in H is its inverse in G.

Proof. If i is the identity element of H, then $ii = i$. But in G, $ie = i$, so $ii = ie$ and, by cancellation, $i = e$.

The proof for inverses is exercise 5. ■

The next theorem is a labor-saving device for proving that a subset of a group is a subgroup. It is given in "iff" form for completeness, but the important result is that only two properties must be checked to show H is a subgroup of G. The first is that H is nonempty. This is usually done by showing that $e \in H$. The other is to show that $ab^{-1} \in H$ whenever a and b are in H. This is usually less work than showing both that H is closed under the group operation and that $b \in H$ implies $b^{-1} \in H$.

Theorem 6.11. Let G be a group. A subset H of G is a group if and only if H is nonempty and for all $a, b \in H, ab^{-1} \in H$.

Proof. First, suppose H is a subgroup of G. Then H is a group, so H contains the identity e. Therefore, $H \neq \varnothing$. Also if a and b are in H, then $b^{-1} \in H$ (by the inverse property) and $ab^{-1} \in H$ (by the closure property).

Now suppose $H \neq \varnothing$ and for all $a, b \in H, ab^{-1} \in H$. ⟨*We show H is a subgroup of G by showing that the group axioms and closure hold for H. It is convenient to proceed in the order which follows.*⟩

(i) Let x, y, and z be in H. Then x, y, and z are in G, so by associativity for G, $(xy)z = x(yz)$.
(ii) Let $a \in H$ (because $H \neq \varnothing$). Then $aa^{-1} = e \in H$.
(iii) Suppose $x \in H$. Then e and x are in H, so by hypothesis, $ex^{-1} = x^{-1} \in H$.
(iv) Let x and y be in H. Then by (iii), $y^{-1} \in H$. Then x and y^{-1} are in H, so by hypothesis $x(y^{-1})^{-1} = xy \in H$. ■

There are subgroups connected with homomorphisms. We know that if f is a homomorphism from $(A, \circ)$ to $(B, *)$, then $(\text{Rng}(f), *)$ is a subgroup of $(B, *)$. This is so because the homomorphic image of a group is a group, and $\text{Rng}(f) \subseteq B$. The next definition and theorem show that for every homomorphism there is a subgroup on the domain side, too.

Definition. Let $f: (A, \circ) \to (B, *)$ be a homomorphism and let i be the identity element of B. The **kernel of f** is $\ker(f) = \{x \in A : f(x) = i\}$.

Theorem 6.12. The kernel of a homomorphism is a subgroup of the domain.

Proof. Let $f: (A, \circ) \to (B, *)$ be a homomorphism and let e and i be the identity elements of A and B, respectively. ⟨*We apply Theorem 6.11.*⟩ First, $\ker(f)$ is nonempty, because $f(e) = i$, by Theorem 6.3; so $e \in \ker(f)$. Suppose a and b are in $\ker(f)$. Then $f(a) = f(b) = i$. By Theorem 6.3, $f(b^{-1}) = (f(b))^{-1}$. Therefore, $f(a \circ b^{-1}) = f(a) \circ f(b^{-1}) =$ $f(a) \circ (f(b))^{-1} = i \circ i^{-1} = i$. This shows that $a \circ b^{-1} \in \ker(f)$. Therefore, $\ker(f)$ is a subgroup of A. ■

There is a particularly nice homomorphism from $(\mathbf{Z}, +)$ onto $(\mathbf{Z}_m, +_m)$ for each m. This is the canonical map defined by $f(a) = a/\equiv_m$ for each $a \in \mathbf{Z}$. By definition of $+_m$, $f(a + b) = (a + b)/\equiv_m = a/\equiv_m +_m b/\equiv_m = f(a) + f(b)$. The kernel of this homomorphism is $0/\equiv_m = \{0, \pm m, \pm 2m, \ldots\}$ which is indeed a subgroup of $\mathbf{Z}$ under addition.

Consider the groups $(\mathbf{Z}_6, +_6)$ and $(\{a, b, c\}, \circ)$ with the operation table for $\circ$ as shown.

$\circ$	a	b	c
a	a	b	c
b	b	c	a
c	c	a	b

It can be verified (exercise 17) by checking all cases that the map $g: \mathbf{Z}_6 \xrightarrow{\text{onto}} \{a, b, c\}$ given by $g(0) = g(3) = a$, $g(1) = g(4) = b$, and $g(2) = g(5) = c$ is a homomorphism. The kernel of g is $\{0, 3\}$, which we know is a subgroup of $\mathbf{Z}_6$.

If a is a member of a group G, then all powers of a are in G by the closure property. This set of all powers of a produces a subgroup of G.

Definitions. Let a be an element of a group G. The **cyclic subgroup generated by a** is $[a] = \{a^n : n \in \mathbf{Z}\}$. If there is $b \in G$ such that $[b] = G$, then G is called a **cyclic group** and b is called a **generator** for G.

The name cyclic subgroup for $[a]$ is justified by the fact that $[a]$ is a subgroup of G. We verify this by observing that $[a]$ is not empty since a^0 is the identity of G. Also, if $x, y \in [a]$, then $x = a^m$ and $y = a^n$ for some integers m and n, so $xy^{-1} = a^m(a^n)^{-1} = a^m a^{-n} = a^{m-n} \in [a]$. Thus $[a]$ is a subgroup of G. Every cyclic group is abelian, because $a^m a^n = a^{m+n} = a^n a^m$.

We have already seen some cyclic groups. The group $(\mathbf{Z}, +)$ is cyclic with generator 1 and every group $(\mathbf{Z}_m, +_m)$ is cyclic with generator $1/\equiv_m$. A cyclic group may have more than one generator. For the group $(\mathbf{Z}_4, +_4)$, $\mathbf{Z}_4 = [1] = [3]$. The element 2 generates the subgroup $\{0, 2\}$. The group S_3 is not cyclic because none of its elements generates the entire group (exercise 16). Of course the fact that S_3 is not abelian is sufficient to conclude that it is not cyclic.

The **order of an element** $a \in G$ is the order of the cyclic subgroup $[a]$ generated by a. If $[a]$ is an infinite set, we say a has infinite order.

Example. In the octic group, the element V has order 2 since $V^0 = I$, $V^1 = V$, $V^2 = VV = I$. Thus, any power of V is either I or V; and therefore, $[V] = \{I, V\}$. Similarly, the element R has order 4 because $[R] = \{I, R, R^2, R^3\}$. The orders of the elements of the octic group are 1 (for I), 2 (for R^2, V, D, and D') and 4 (for R and R^3). Thus the octic group is not cyclic.

Theorem 6.13. Let G be a group and a be an element of G with order r. Then r is the smallest positive integer such that $a^r = e$, the identity, and $[a] = \{e, a, a^2, \ldots, a^{r-1}\}$.

Proof. Since $[a]$ is finite the powers of a are not all distinct. Let $a^m = a^n$ with $0 \leq m < n$. Then $a^{n-m} = e$ with $n - m > 0$. Therefore, the set of positive integers p such that $a^p = e$ is nonempty. Let k be the smallest such integer. ⟨*This k exists by the Well-Ordering Principle.*⟩ We prove that $k = r$ by showing that the elements of $[a]$ are exactly $a^0 = e, a^1, a^2, \ldots,$ and a^{k-1}.

First, the elements $e, a^1, a^2, \ldots, a^{k-1}$ are distinct, for if $a^s = a^t$ with $0 \leq s < t < k$, then $a^{t-s} = e$ and $0 < t - s < k$, contradicting the definition of k.

Second, every element of $[a]$ is one of $e, a^1, a^2, \ldots, a^{k-1}$. Consider a^t for $t \in \mathbf{Z}$. By the division algorithm, $t = mk + s$ with $0 \leq s < k$. Thus $a^t = a^{mk+s} = a^{mk}a^s = (a^k)^m a^s = e^m a^s = ea^s = a^s$, so that $a^t = a^s$ with $0 \leq s < k$.

We have shown that the elements a^s for $0 \leq s < k$ are all distinct and that every power of a is equal to one of these. Since $[a]$ has exactly r elements, $r = k$ and $a^r = e$. ■

If $a \in G$ has infinite order, then by the reasoning used in the proof of Theorem 6.13, all the powers of a are distinct and $[a] = \{\ldots, a^{-2}, a^{-1}, a^0 = e, a^1, a^2, \ldots\}$.

Because homomorphisms preserve structure, you should not be surprised that a homomorphic image of a cyclic group is cyclic. In fact, the image of a generator of the group is a generator of the image group. Furthermore, every subgroup of a cyclic group is cyclic. This means that if $G = [a]$ and H is a subgroup of G, then $H = [a^s]$ for some $s \in \mathbf{N}$. The proofs are exercises 20 and 21.

Exercises 6.4

☆ 1. Find all subgroups of
(a) $(\mathbf{Z}_6, +)$ (b) $(\mathbf{Z}_8, +)$ (c) $(\mathbf{Z}_7 - \{0\}, \cdot)$

2. Find all subgroups of the following group. (There are six.)

	a	b	c	d	e	f
a	a	b	c	d	e	f
b	b	a	f	e	d	c
c	c	e	a	f	b	d
d	d	f	e	a	c	b
e	e	c	d	b	f	a
f	f	d	b	c	a	e

3. (a) Find all subgroups of S_3.
   (b) Find a homomorphism from the group in exercise 2 to S_3 that is one-to-one and onto S_3.
4. For the octic group of symmetries of a square in section 6.3 give the five different subgroups of order two and the three different subgroups of order four.
5. Prove that, if G is a group and H is a subgroup of G, then the inverse of an element $x \in H$ is the same as its inverse in G. (Theorem 6.10.)
6. Prove that if H and K are subgroups of a group G, then $H \cap K$ is a subgroup of G.
7. Prove that if $\{H_\alpha : \alpha \in \Delta\}$ is a family of subgroups of a group G, then $\bigcap_{\alpha \in \Delta} H_\alpha$ is a subgroup of G.
8. Give an example of a group G and subgroups H and K of G such that $H \cup K$ is not a subgroup of G.

★ 9. Let G be a group with identity e and let $a \in G$. Prove that the set $N_a = \{x \in G : xa = ax\}$, called the **normalizer** of a in G, is a subgroup of G.

10. Let G be a group. Prove that the **center** of G, $C = \{x \in G$: for all $y \in G$, $xy = yx\}$, is a subgroup of G.
11. Prove that if G is a group and $a \in G$, then the center of G is a subgroup of the normalizer of a in G.
12. Is there a subgroup H of S_3 such that (312) is in H but (231) is not? Explain.
13. Let G be a group and H be a subgroup of G.
    (a) If G is abelian must H be abelian? Explain.
    (b) If H is abelian must G be abelian? Explain.
14. Let G be a group. If H is a subgroup of G and K is a subgroup of H, prove that K is a subgroup of G.

★ 15. Let G be a group and let H be a subgroup of G. Let a be a fixed element of G. Prove that $K = \{a^{-1}ha : h \in H\}$ is a subgroup of G.

16. For each element of S_3 find the cyclic subgroup of S_3 generated by that element.
17. Let $(\{a, b, c\}, \circ)$ is the group with operation table

$\circ$	a	b	c
a	a	b	c
b	b	c	a
c	c	a	b

Verify that the mapping $g: (\mathbf{Z}_6, +) \to (\{a, b, c\}, \circ)$ defined by $g(0) = g(3) = a$, $g(1) = g(4) = b$, and $g(2) = g(5) = c$ is a homomorphism.

18. Let $(\mathbf{C} - \{0\}, \cdot)$ be the group of complex numbers with ordinary complex number multiplication. Let $\alpha = (1 + i\sqrt{3})/2$.
    (a) Find $[\alpha]$.
    (b) Find a generator of $[\alpha]$ other than α.
19. Let x be an element of the group G with identity e. What are the possibilities for the order of x if
    ★ (a) $x^{15} = e$
    (b) $x^{20} = e$
    (c) $x^n = e$.
20. Prove that every subgroup of a cyclic group is cyclic.
21. Let $h: G \to K$ be a homomorphism.
    (a) Prove that if $x \in G$, then $h(x^k) = (h(x))^k$ for all $k \in \mathbf{N}$.
    (b) Prove that the homomorphic image of a cyclic group is cyclic.
22. Let $G = [a]$ be a cyclic group of order 30.
    (a) What is the order of a^6?
    (b) List all elements of order 2.
    ★ (c) List all elements of order 3.
    (d) List all elements of order 10.

23. Let $G = [a]$ be a cyclic group of order 9. Let G' be the group generated by the permutation (231). If Φ is a homomorphism from G to G' such that $\Phi(a) = (231)$, find all other images.
24. Let $\Phi : (G, *) \to (H, \circ)$ be a homomorphism. Let e be the identity of G. Prove that Φ is one-to-one iff $\ker(\Phi) = \{e\}$.
☆ 25. Let $f: G \to H$ be a homomorphism. For $a \in G$, prove the order of $f(a)$ divides the order of a.

SECTION 6.5. COSETS AND LAGRANGE'S THEOREM

The purpose of this section is to relate the order of a subgroup of a given group to the order of the group. We shall see that any subgroup of a group may be used to partition the group into a collection of subsets called cosets.

Definition. Let $(G, \circ)$ be a group, H a subgroup of G, and $x \in G$. The set $x \circ H = \{x \circ h : h \in H\}$ is called the **left coset** of x and H and $H \circ x = \{h \circ x : h \in H\}$ is called the **right coset** of x and H.

Table 6.7 below shows a group T with identity a and subgroups $H = \{a, b\}$ and $K = \{a, e, f\}$. The cosets of H and K are

TABLE 6.7 Group T

	a	b	c	d	e	f
a	a	b	c	d	e	f
b	b	a	f	e	d	c
c	c	e	a	f	b	d
d	d	f	e	a	c	b
e	e	c	d	b	f	a
f	f	d	b	c	a	e

Cosets of H
$aH = \{a, b\} = Ha = \{a, b\}$
$bH = \{b, a\} = Hb = \{b, a\}$
$cH = \{c, e\} \neq Hc = \{c, f\}$
$dH = \{d, f\} \neq Hd = \{d, e\}$
$eH = \{e, c\} \neq He = \{e, d\}$
$fH = \{f, d\} \neq Hf = \{f, c\}$

Cosets of K
$aK = \{a, e, f\} = Ka = \{a, e, f\}$
$bK = \{b, d, c\} = Kb = \{b, c, d\}$
$cK = \{c, b, d\} = Kc = \{c, d, b\}$
$dK = \{d, c, b\} = Kd = \{d, b, c\}$
$eK = \{e, f, a\} = Ke = \{e, f, a\}$
$fK = \{f, a, e\} = Kf = \{f, a, e\}$

Notice that the left coset fH and the right coset Hf are not equal. It does happen that for every x in T, $xK = Kx$.

As another example of cosets, consider the subgroup $H = \{0, \pm 5, \pm 10, \ldots\}$ of $(\mathbf{Z}, +)$. Then $2 + H = H + 2 = \{\ldots, -8, -3, 2, 7, \ldots\} = 2/\equiv_5$ and $0 + H = H + 0 = H$.

Theorem 6.14. If H is a subgroup of G and $x \in G$, then H and xH are equivalent.

Proof. To show H and xH are equivalent, we show that the function F given by $F(h) = xh$ maps H one-to-one and onto xH. If $w \in xH$, then $w = xh$ for some $h \in H$. Then for that h, $F(h) = w$. Therefore, F maps onto xH. Suppose $F(h_1) = F(h_2)$ for $h_1, h_2 \in H$. Then $xh_1 = xh_2$. Then, by cancellation, $h_1 = h_2$. Therefore, F is one-to-one. ■

Because every left coset of H is equivalent to H, any two left cosets of H have the same number of elements. This result and others to follow also apply to right cosets.

Theorem 6.15. The set of left cosets of H in G is a partition of G.

Proof. Every element x of G is a member of some left coset because $x = xe \in xH$. Therefore, G is a union of left cosets. It remains to show that left cosets are either identical or disjoint.

Suppose $xH \cap yH \neq \emptyset$. Let $b \in xH \cap yH$. Then $b = xh_1 = yh_2$ for some $h_1, h_2 \in H$. Therefore, $x = yh_2h_1^{-1}$ and $xH = yh_2h_1^{-1}H$. Now let xh be in xH. Then $xh = yh_2h_1^{-1}h'$ for some $h' \in H$. But $h_2h_1^{-1}h' \in H$, so $xh \in yH$. This shows that $xH \subseteq yH$ and a similar argument shows $yH \subseteq xH$. Therefore, $xH = yH$. ■

Theorem 6.16. (Lagrange). Let G be a finite group and H a subgroup of G. Then the order of H divides the order of G.

Proof. Let the order of G be n and the order of H be m. Let the number of left cosets of H in G be k. Denote the distinct cosets of H by $x_1H, x_2H, \ldots, x_kH$. By Theorem 6.14, each coset contains exactly m elements. By Theorem 6.15, $G = x_1H \cup x_2H \cup \ldots \cup x_kH$, and the cosets are pairwise disjoint. Thus there are mk elements in G. Therefore, $n = mk$, and so m divides n. ■

For a group G with order n and subgroup H of order m, the integer n/m is called the **index** of H in G and is the number of distinct left cosets of H.

The converse of Lagrange's Theorem is false. There exists a group of order 12 which has no subgroup of order 6.

Corollary 6.17. Let G be a finite group and $a \in G$. Then the order of a divides the order of G.

Proof. Exercise 6. ■

Corollary 6.18. If G is a finite group of order n and $a \in G$, then $a^n = e$, the identity.

Proof. Exercise 7. ■

Corollary 6.19. A group of prime order is cyclic.

Proof. Let G have prime order p. Let a be an element of G other than e. By Corollary 6.17, the order of a divides the order of G, and so must be either 1 or p. But $[a]$ contains both e and a, so the order of a must be p. Therefore, $G = [a]$ and G is cyclic. ■

Exercises 6.5

☆ 1. Let H be the subgroup of S_3 generated by the permutation (213). Find all left and right cosets of H in S_3.

2. Let H be the subgroup of S_3 generated by the permutation (231). Find all left and right cosets of H in S_3.

3. Find all the cosets of $H = [a^3]$ in a cyclic group $G = [a]$ of order 12.

★ 4. Show that a group G of order 4 is either cyclic or is such that $x^2 = e$ for all $x \in G$.

5. Show that a group G of order 27 is either cyclic or is such that $x^9 = e$ for all $x \in G$.

6. Prove Corollary 6.17.

7. Prove Corollary 6.18.

8. Let H be a subgroup of G.
   (a) Prove that $aH = bH$ iff $a^{-1}b \in H$.
   (b) If $aH = bH$, must $b^{-1}a$ be in H? Explain.
   (c) If $aH = bH$, must ab^{-1} be in H? Explain.

☆ 9. Prove that if G is a cyclic group of order n and if m divides n, then G has a subgroup of order m.

10. **Proofs to Grade.**

★ (a) **Claim.** If H is a subgroup of G and $aH = bH$, then $a^{-1}b \in H$.
"Proof." Suppose $aH = bH$. Then $ah = bh$ for some $h \in H$, and so $h = a^{-1}bh$. Thus $hh^{-1} = a^{-1}b$ and since $hh^{-1} \in H$, we have that $a^{-1}b \in H$. ■

(b) **Claim.** If H is a subgroup of G and $a^{-1}b \in H$, then $aH \subseteq bH$.
"Proof." Suppose $a^{-1}b \in H$. Then $a^{-1}b = h$ for some $h \in H$. Thus $a^{-1} = hb^{-1}$, so that $a = (hb^{-1})^{-1} = bh^{-1}$. Thus $a \in bH$. Now, if $x \in aH$, then $x = ah'$ for some $h' \in H$. Thus $x = (bh^{-1})h' = b(h^{-1}h')$, so $x \in bH$. Therefore, $aH \subseteq bH$. ■

★ (c) **Claim.** If H is a subgroup of G and $aH = bH$, then $b^{-1}a \in H$.
"Proof." Suppose $aH = bH$. Then $b^{-1}aH = b^{-1}bH = eH = H$. Therefore $b^{-1}aH = H$. Since $b^{-1}a \in b^{-1}aH$, $b^{-1}a \in H$. ■

SECTION 6.6. QUOTIENT GROUPS

Let G be a group and let H be a subgroup of G. The set of left cosets of H in G is a partition of G, by Theorem 6.15. Therefore, there is a corresponding equivalence relation on G with the property that $a, b \in G$ are related iff they belong to the same left coset of H. (See chapter 3, section 3.) The equivalence classes under this relation are exactly the left cosets of H. The set of all equivalence classes (cosets) is denoted G/H (read "G mod H").

Example. In $(\mathbf{Z}_6, +)$, let H be the subgroup $\{0, 3\}$. The left cosets of H are

$$0 + H = 3 + H = \{0, 3\}$$
$$1 + H = 4 + H = \{1, 4\}$$

and

$$2 + H = 5 + H = \{2, 5\}.$$

Here $\mathbf{Z}_6 / H = \{0 + H, 1 + H, 2 + H\}$.

It is our intention to impose a structure—an operation we hope will satisfy the group axioms—on G/H. If we try to define a "product" for cosets xH and yH, nothing could be more natural than $xH \cdot yH = xyH$. In order to try out this "structure" $(G/H, \cdot)$ we consider as examples the sets T/H and T/K where T, H, and K are the groups presented in table 6.7.

The left cosets of H in T are $\{a, b\}$, $\{c, e\}$, and $\{d, f\}$. Then, applying the definition for $\cdot$ on T/H, we have

$$\{c, e\} \cdot \{d, f\} = cH \cdot dH = cdH = fH = \{d, f\}$$

and

$$\{c, e\} \cdot \{d, f\} = eH \cdot fH = efH = aH = \{a, b\}.$$

The fact that $\cdot$ yields two different images means that it is not an operation on T/H.

The left cosets of K in T are $\{a, e, f\} = aK = eK = fK$ and $\{b, c, d\} = bK = cK = dK$. In this case $\cdot$ is indeed an operation on T/K, because the product $xK \cdot yK$ does not depend on the representatives for the cosets. For example, $eK = fK$, and $bK = cK$, and $eK \cdot bK = ebK = cK$, and $fK \cdot cK = fcK = bK$, but $cK = bK$. Table 6.8 is the operation table for T/K.

TABLE 6.8

	aK	bK
aK	aK	bK
bK	bK	aK

That the proposed operation $\cdot$ works for T/K and $xK = Kx$ for all $x \in T$ is more than coincidence. We are led to the following definition.

Definition. Let G be a group and H a subgroup of G. Then H is called **normal** in G iff for all $x \in G$, $xH = Hx$.

Thus K is a normal subgroup of T but H is not. If G is any group, then both G and $\{e\}$ are normal in G (see exercise 4). If G is an abelian group, then every subgroup of G is normal. Notice that $xK = Kx$ does not imply that $xk = kx$ for every $k \in K$. For the subgroup K of T considered above, $f \in K$ and $dK = Kd$, but $df \neq fd$. (See table 6.7.)

Theorem 6.20. If H is a normal subgroup of G with $xH = yH$ and $wH = vH$, then $xwH = yvH$.

Proof. Let xwh be an element of xwH. Then $wh \in wH = vH = Hv$ ⟨*because H is normal*⟩, so $wh = h_1v$ for some $h_1 \in H$. Thus $xwh = xh_1v$. Now $xh_1 \in xH = yH = Hy$, so $xh_1 = h_2y$ for some $h_2 \in H$. Thus $xwh = h_2yv \in Hyv = yvH$. Since cosets are either identical or disjoint, $xwH = yvH$. ■

If H is a normal subgroup of G and if $xH = yH$ and $wH = vH$, then $xH \cdot wH = xwH = yvH = yH \cdot vH$, by Theorem 6.20. This establishes that $(G/H, \cdot)$ is an algebraic structure.

Theorem 6.21. If H is a normal subgroup of G, then $(G/H, \cdot)$, with $\cdot$ as defined above, is a group called **the quotient group of G modulo H** (G mod H). The identity element is the coset $H = eH$ and the inverse of xH is $x^{-1}H$.

Proof. Exercise 3. ■

Let $m \in \mathbf{N}$ and let H_m be the subgroup of $(\mathbf{Z}, +)$ consisting of all multiples of m. Then H_m is normal because $(\mathbf{Z}, +)$ is abelian. Elements of $\mathbf{Z}/H_m$ are cosets of the form $x + H_m$, which is the equivalence class $x/\equiv_m$. The operation on cosets in $\mathbf{Z}/H_m$ is the same as the operation $+_m$ on $\mathbf{Z}_m$. Therefore, $\mathbf{Z}/H_m$ is the group $(\mathbf{Z}_m, +_m)$.

The group $H = (\{1, 6\}, \cdot_7)$ is a normal subgroup of the abelian group $(\mathbf{Z}_7 - \{0\}, \cdot_7)$. The cosets in $(\mathbf{Z}_7 - \{0\})/\{1, 6\}$ are H, $2H = \{2, 5\}$, and $3H = \{3, 4\}$. Table 6.9 shows $(\mathbf{Z}_7 - \{0\}, \cdot_7)$ and the quotient group.

TABLE 6.9 $(\mathbf{Z}_7 - \{0\}, \cdot_7)$

$\cdot_7$	1	2	3	4	5	6
1	1	2	3	4	5	6
2	2	4	6	1	3	5
3	3	6	2	5	1	4
4	4	1	5	2	6	3
5	5	3	1	6	4	2
6	6	5	4	3	2	1

$((\mathbf{Z}_7 - \{0\})/\{1, 6\}, \cdot)$

$\cdot$	H	$2H$	$3H$
H	H	$2H$	$3H$
$2H$	$2H$	$3H$	H
$3H$	$3H$	H	$2H$

Exercises 6.6

★ 1. Let G be a group and H be a subgroup of G. Prove that H is normal iff $a^{-1}Ha \subseteq H$ for all $a \in G$, where $a^{-1}Ha = \{a^{-1}ha : h \in H\}$.

☆ 2. Prove that a subgroup H of G is normal in G iff $Ha \subseteq aH$ for every $a \in G$.

3. Prove Theorem 6.21.

4. Let G be a group with identity e.
   (a) Prove that $\{e\}$ and G are normal subgroups of G.
   (b) Describe the quotient group $G/\{e\}$.
   (c) Describe the quotient group G/G.

5. Let H be a normal subgroup of G. Prove that
   (a) if G is abelian, then G/H is abelian.
   ☆ (b) if G is cyclic, then G/H is cyclic.

6. Let G be the group $(\mathbb{R} - \{0\}, \cdot)$ and let H be the subgroup consisting of all positive real numbers. Show that H is a normal subgroup of G of index 2 and construct the operation table for G/H.

7. For the group $\mathcal{O}$ of symmetries of a square (section 6.3), let $J = \{I, R, R^2, R^3\}$, $K = \{I, R^2, H, V\}$, and $L = \{I, H\}$.
   (a) Construct the operation table for $\mathcal{O}/J$.
   (b) Construct the operation table for $\mathcal{O}/K$.
   (c) Is L normal in $\mathcal{O}$?
   (d) Is L normal in K?

☆ 8. If H_1 is a normal subgroup of G, and H_2 is a normal subgroup of H_1, must H_2 be a normal subgroup of G? Explain.

9. Consider the subgroup $C = \{I, R^2\}$ of the octic group $\mathcal{O}$.
   (a) Prove that C is normal in the octic group and hence that the quotient group $\mathcal{O}/C$ exists.
   (b) Construct the operation table for $\mathcal{O}/C$.
   (c) Is $\mathcal{O}/C$ abelian?
   (d) Is $\mathcal{O}/C$ cyclic?

10. Let f be a homomorphism of G onto H. Let N be a normal subgroup of G. Prove that $f(N) = \{f(n) : n \in N\}$ is a normal subgroup of H.

11. Let f be a homomorphism of G onto H. Let M be a subgroup of H. Prove that $N = \{g : f(g) \in M\}$ is a subgroup of G that contains $\ker(f)$. Show that if M is normal in H, then N is normal in G.

12. **Proofs to Grade.**
   (a) **Claim.** If H is a normal subgroup of G and K is a normal subgroup of H, then K is a normal subgroup of G.
   "Proof." We must show $xK = Kx$ for all $x \in G$. Since H is normal in G, $xH = Hx$ for all $x \in G$. Since K is a normal subgroup of H, $xK = Kx$ for all $x \in H$ and hence for all $x \in G$. Therefore, K is a normal subgroup of G. ■
   ★ (b) **Claim.** If H and K are normal subgroups of G, then $H \cap K$ is a normal subgroup of G.
   "Proof." Let $x \in G$. Then $x(H \cap K) = xH \cap xK = Hx \cap Kx$ ⟨*since H and K are normal*⟩ $= (H \cap K)x$. Therefore, $H \cap K$ is normal. ■
   (c) **Claim.** If H is a normal subgroup of G and G/H is abelian, then G is abelian.
   "Proof." Let $x, y \in G$. Then $xH, yH \in G/H$. Since G/H is abelian $(xH)(yH) = (yH)(xH)$. Therefore $xyH = yxH$. Therefore $(xy)h = (yx)h$ for some $h \in H$. By cancellation, $xy = yx$. ■

SECTION 6.7. ISOMORPHISM; THE FUNDAMENTAL THEOREM OF GROUP HOMOMORPHISMS

A homomorphism that is one-to-one is called an **isomorphism.** If there is an isomorphism $f: (A, \circ) \xrightarrow{\text{onto}} (B, *)$, then $(A, \circ)$ is said to be **isomorphic** to $(B, *)$. Inverses and composites of isomorphisms are also isomorphisms, so the relation of being isomorphic is an equivalence relation on the class of all groups.

Isomorphic groups are literally "same-structured" because they differ only in the names or nature of their elements; all their algebraic properties are identical. For example, the groups $(\mathbf{Z}_2, +_2)$, $(\{1, 6\}, \cdot_7)$ and $(\{\varnothing, A\}, \Delta)$ are isomorphic, where A is any nonempty set and Δ is the symmetric difference operation defined by $X \Delta Y = (X - Y) \cup (Y - X)$. The three groups are shown in table 6.10.

TABLE 6.10

$+_2$	0	1
0	0	1
1	1	0

$\cdot_7$	1	6
1	1	6
6	6	1

Δ	$\varnothing$	A
$\varnothing$	$\varnothing$	A
A	A	$\varnothing$

In fact, any two groups of order 2 are isomorphic. This observation can be extended: every cyclic group of order m is isomorphic to $(\mathbf{Z}_m, +_m)$. Every infinite cyclic group is isomorphic to $(\mathbf{Z}, +)$ (see exercises 6–9). Because of these results, we can say that $(\mathbf{Z}_m, +_m)$ and $(\mathbf{Z}, +)$ are the only cyclic groups, up to isomorphism. The noncyclic group T of section 6.5 is isomorphic to the group $(S_3, \circ)$ under the isomorphism
$f = \{(a, (123)), (b, (213)), (c, (321)), (d, (132)), (e, (231)), (f, (312))\}$.

We claimed in the introduction to section 6.3 that for every group there is a permutation group with the same structure. We now prove this result, due to Arthur Cayley (1821–1895).

Theorem 6.22. Every group G is isomorphic to a permutation group.

Proof. As the set whose elements are to be permuted we choose the set G. If $a \in G$, then $\{ax: x \in G\} = G$. Therefore, the function $\theta_a: G \to G$ defined by $\theta_a(x) = ax$ is a mapping from G onto G which ⟨*by the cancellation property of the group*⟩ is one-to-one. Therefore, θ_a is a permutation of G associated with the element a of G. The function f sending a to θ_a will be our isomorphism from G to the set $H = \{\theta_a: a \in G\}$.

Let $x \in G$. Then $\theta_{ab}(x) = (ab)x = a(bx) = \theta_a(\theta_b(x)) = (\theta_a \circ \theta_b)(x)$, so $\theta_{ab} = \theta_a \circ \theta_b$. Therefore, f is a homomorphism. Also, f maps onto H by definition of H. Thus H ⟨*the homomorphic image*⟩ is a permutation group. Now suppose $\theta_a = \theta_b$. Then $ax = bx$ for every $x \in G$, so that $a = b$. Therefore, f is one-to-one; G and H are isomorphic. ■

The main purpose of this section is to establish the connection between homomorphisms defined on a group and normal subgroups. First, for every homomorphism, there is a corresponding normal subgroup. This subgroup determines a quotient group which is isomorphic to the image group.

Theorem 6.23 (a). The kernel K of a homomorphism f from a group $(G, \circ)$ onto $(B, *)$ is a normal subgroup of G. Furthermore, $(G/K, \cdot)$ is isomorphic to $(B, *)$.

Proof. We saw in Theorem 6.12 that $K = \{y \in G: f(y) = e_B\}$ ⟨*where e_B is the identity of B*⟩ is a subgroup of G. We first prove that for all x and $a \in G$, $x \in a \circ K$ iff $f(x) = f(a)$. Suppose $x \in a \circ K$. Then $x = a \circ k$ for some $k \in K$. Then $f(x) = f(a \circ k) = f(a) * f(k) = f(a) * e_B = f(a)$, so $f(x) = f(a)$. Now suppose $f(x) = f(a)$. Then $f(a^{-1} \circ x) = f(a^{-1}) * f(x) = (f(a))^{-1} * f(a) = e_B$. Thus $a^{-1} \circ x \in K$. Therefore, $x = a \circ (a^{-1} \circ x) \in aK$.

Similarly, $x \in K \circ a$ iff $f(x) = f(a)$. Hence, $x \in a \circ K$ iff $f(x) = f(a)$ iff $x \in K \circ a$, so $a \circ K = K \circ a$ for all $a \in G$. Therefore, K is normal in G.

The argument above shows that for every left coset $a \circ K$ of the kernel K, all the elements of $a \circ K$ are mapped by f to the same element $f(a)$ in B. Therefore, we may define a mapping $h: G/K \to B$ by setting $h(a \circ K) = f(a)$. Furthermore, h is one-to-one. ⟨*If $h(a_1 \circ K) = h(a_2 \circ K)$, then $f(a_1) = f(a_2)$, so $a_2 \in a_1 \circ K$. As $a_2 = a_2 \circ e$ is also in $a_2 \circ K$, and cosets form a partition of G, $a_1 \circ K = a_2 \circ K$.*⟩

The mapping h is onto B: if $b \in B$, then $b = f(a)$ for some $a \in G$, and $b = f(a) = h(a \circ K)$.

Finally, h is operation preserving: $h((a_1 \circ K)(a_2 \circ K)) = h(a_1 \circ a_2 \circ K) = f(a_1 \circ a_2) = f(a_1) * f(a_2) = h(a_1 \circ K) * h(a_2 \circ K)$. Therefore, h is an isomorphism from $(G/K, \cdot)$ onto $(B, *)$. ■

The other aspect of the connection between homomorphisms and normal subgroups is that it is possible to start with any normal subgroup and construct a homomorphic image of the group. This image is the quotient group.

Theorem 6.23 (b). Let H be a normal subgroup of $(G, \circ)$. Then the mapping $g: G \to G/H$ given by $g(a) = a \circ H$ is a homomorphism from G onto G/H with kernel H.

Proof. We have seen the mapping g before (see. p. 72) and called it the canonical map from G to the set of cosets ⟨*equivalence classes*⟩. First, g maps onto G/H because, if $a \circ K$ is a left coset, then $a \circ K = g(a)$. To verify that g is operation preserving we compute $g(a \circ b) = a \circ b \circ H = a \circ H \cdot b \circ H = g(a) \cdot g(b)$.

It remains to show that H is the kernel of g. The identity of $(G/H, \cdot)$ is the coset H, so $x \in \ker(g)$ iff $g(x) = x \circ H = H$ iff $x \in H$. Therefore, $\ker(g) = H$. ■

The two theorems above are summarized in:

Theorem 6.24. (The Fundamental Theorem of Group Homomorphisms). Every homomorphism of a group G determines a normal subgroup of G. Every normal subgroup of G determines a homomorphic image of G. Up to isomorphism the only homomorphic images of G are the quotient groups of G modulo normal subgroups. Finally, the diagram in figure 6.4 "commutes"; that is, $f = h \circ g$, where h and g are the mappings of Theorems 6.23 (a) and (b).

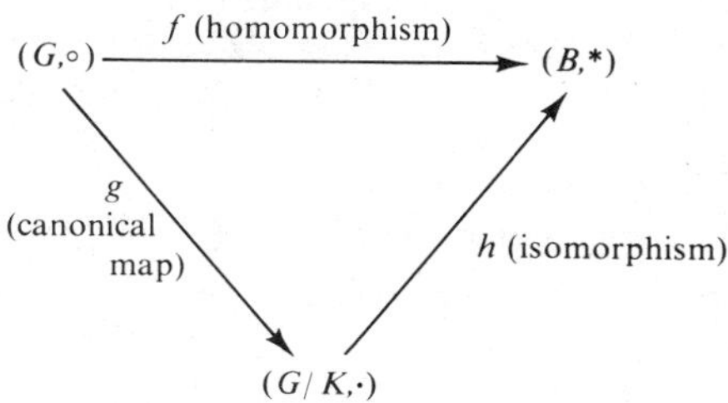

Figure 6.4

Proof. What remains to be proved is that f is the composite of h with g. Let $a \in G$. Then $g(a) = a \circ K$ and $h(g(a)) = h(a \circ K) = f(a)$. Therefore, $f = h \circ g$. ■

Finally, we note that the theorem's primary assertion is that every homomorphic image can be identified with (is isomorphic to) the quotient group $G/\ker(f)$. Furthermore, with this identification, the homomorphism f acts like the canonical homomorphism. Thus, we can think of homomorphic images as though they are quotient groups, and homomorphisms as though they are canonical maps. The Fundamental Theorem is used to conclude that two groups are isomorphic to each other without actually constructing the isomorphism.

As an example of the use of the Fundamental Theorem, we describe all homomorphic images of $(\mathbf{Z}_6, +)$. Every such image must be isomorphic to a quotient group of $\mathbf{Z}_6$. The subgroups of $\mathbf{Z}_6$ have orders 1, 2, 3, and 6 and every subgroup is normal in $\mathbf{Z}_6$. The corresponding quotient groups have orders 6, 3, 2, and 1. These quotient groups are cyclic (see exercise 5(b) of section 6.6) and therefore isomorphic to $\mathbf{Z}_6$, $\mathbf{Z}_3$, $\mathbf{Z}_2$, and $\{0\}$, respectively.

Exercises 6.7

1. Show that the mapping $f: (\mathbf{R}^+, \cdot) \to (\mathbf{R}, +)$ given by $f(x) = \log_{10} x$ is an isomorphism.
2. Is $(\mathbf{Z}_4, +)$ isomorphic to $(\{1, -1, i, -i\}, \cdot)$? Explain.
3. Is S_3 isomorphic to $(\mathbf{Z}_6, +)$? Explain.
4. Give an example of two groups of order 8 that are not isomorphic.
5. ☆ Let $f: (G, \cdot) \to (H, *)$ be an isomorphism. For $a \in G$, prove the order of a equals the order of $f(a)$.
6. Prove that any two groups of order 2 are isomorphic.

7. Prove that any two groups of order 3 are isomorphic.
8. Show that every cyclic group of order m is isomorphic to $(\mathbf{Z}_m, +)$.
9. Show that every infinite cyclic group is isomorphic to $(\mathbf{Z}, +)$.
10. Use the method of proof of Cayley's Theorem to find a group of permutations isomorphic to
    (a) $(\mathbf{Z}_5 - \{0\}, \cdot_5)$
    (b) the subgroup $\{I, R^2, D, D'\}$ of the octic group
    (c) $(\mathbf{R}, +)$
11. (a) Define a homomorphism from $(\mathbf{Z}_{2n}, +)$ onto $(\mathbf{Z}_n, +)$.
    (b) What is the kernel K of the homomorphism you defined in (a)?
    (c) Prove that $\mathbf{Z}_{2n} / K$ is isomorphic to $\mathbf{Z}_n$.
    (d) Construct the operation tables for the groups described in (a), (b), and (c) in the case $n = 3$.
12. Describe all homomorphic images of $\mathbf{Z}_8$.
13. Describe all homomorphic images of $\mathbf{Z}_{11}$.
14. Describe all homomorphic images of S_3.

Real Analysis 7

In this chapter we give an introduction to the analyst's point of view of the real numbers. Section 7.1 discusses the reals as an example of a complete ordered field. Sections 7.2 and 7.3 deal with topological properties of $\mathbf{R}$. Section 7.4 is about sequences of real numbers, while section 7.5 ties all the sections together.

SECTION 7.1. FIELD PROPERTIES OF THE REAL NUMBERS

In this section we shall describe the real numbers as a complete ordered field, making precise what these words mean. We shall assume a knowledge of rational, real, and complex number systems commonly obtained in elementary calculus.

Definition. A **field** is an algebraic structure $(F, +, \cdot)$ where $+$ and $\cdot$ are binary operations on F such that

(1) $(F, +)$ is an abelian group with identity denoted by 0.
(2) $(F - \{0\}, \cdot)$ is an abelian group with identity denoted by 1.
(3) For all a, b, c in F, $a \cdot (b + c) = a \cdot b + a \cdot c$.
(4) $0 \neq 1$.

The rationals, the reals, and the complex numbers are all fields. Some examples of finite fields are $(\mathbf{Z}_p, +, \cdot)$ where p is a prime number (see section 6.3). By property (4) every field must have at least two elements and thus the smallest field is $\mathbf{Z}_2 = \{0, 1\}$ with operations

+	0	1
0	0	1
1	1	0

·	0	1
0	0	0
1	0	1

At this point we have no way to describe algebraic properties of one field that distinguish it from others. The next definition gives us a property of the fields **Q** and **R** not shared by the other fields we have named. It is **R** we want to study.

> **Definition.** A field $(F, +, \cdot)$ is **ordered** iff there is a relation $<$ on F such that for all $x, y, z \in F$
>
> (5) $x \not< x$ (irreflexivity).
> (6) If $x < y$ and $y < z$, then $x < z$ (transitivity).
> (7) Either $x < y$, $x = y$, or $y < x$ (trichotomy).
> (8) If $x < y$, then $x + z < y + z$.
> (9) If $x < y$ and $0 < z$, then $xz < yz$.

Properties (8) and (9) insure the compatibility of the order relation $<$ with the field operations $+$ and $\cdot$. Familiar properties such as $0 < 1$, $-1 < 0$, and if $a < b$ then $-b < -a$ can be derived from the order axioms. The notation $x \leq y$ means $x < y$ or $x = y$, while $x > y$ stands for $y < x$.

Both the rational numbers and the real numbers are ordered fields under their usual orderings. It can be shown that the complex numbers do not have an ordering satisfying properties (5)–(9). Consider for example the situation if we assume $0 < i$. Then $0 = 0 \cdot i < i \cdot i = i^2 = -1$ and thus $0 < -1$. On the other hand, if $i < 0$, then $i^2 > 0 \cdot i$, so again $-1 > 0$. These contradictions prevent the complex numbers from being ordered.

The finite fields also cannot be ordered. To show, for example, that $\mathbf{Z}_2$ is not ordered, suppose $0 < 1$. Then $1 = 0 + 1 < 1 + 1 = 0$, so $1 < 0$. Thus, by transitivity, $0 < 0$, contradicting property (5).

We now have a way to distinguish between **R** and **C**. To develop a distinction between **R** and **Q** we first note that every ordered field has the property that between any two elements there is a third element. If $a < b$ is an ordered field, then, using $\frac{1}{2}$ as the inverse of the element $1 + 1$, we have $a < \frac{1}{2}(a + b) < b$. Thus ordered fields do not have "gaps."

However, some ordered fields, like the rationals, seem to be missing some points. For instance, if we look at a sequence of rationals such as 1.4, 1.41, 1.414, 1.4142, 1.41421, 1.414213, ... formed by approximating $\sqrt{2}$, we see the sequence is contained entirely in the rationals, yet the value it approaches ($\sqrt{2}$) is not a rational number. In this respect the rationals appear to be incomplete, even though the system of rational numbers has no gaps. We will now make the idea of being incomplete or having missing points precise.

> **Definition.** Let A be a subset of an ordered field F. We say $u \in F$ is an **upper bound** for A iff $a \leq u$ for all $a \in A$. If A has an upper bound, A is **bounded above.** Likewise, $\ell \in F$ is a **lower bound** for A iff $\ell \leq a$ for all $a \in A$, and A is **bounded below** iff any lower bound for A exists. The set A is **bounded** iff A is both bounded above and bounded below.

In **R**, the half-open interval $[0, 3)$ has 3 as an upper bound. In fact π, 18, and 206 are also upper bounds for $[0, 3)$. Both $-.5$ and 0 are lower bounds. We note that a bound for a set might or might not be an element of the set. The set **N** is bounded below but not above.

In **Q**, the set $A = \{x \in \mathbf{Q} : x < \sqrt{2}\}$ has many upper bounds: 8, 1.42, 1.4146, and so on. A has no lower bounds.

The best possible upper bound for a set A is called the supremum of A.

> **Definition.** Let A be a subset of an ordered field F. We say $s \in F$ is the **least upper bound** or **supremum of *A* in *F*** iff
>
> (1) s is an upper bound for A.
> (2) $s \leq t$ for all upper bounds t of A.
>
> Likewise $i \in F$ is the **greatest lower bound** or **infimum of *A* in *F*** iff
>
> (1) i is a lower bound for A.
> (2) $\ell \leq i$ for all lower bounds ℓ of A.

While several numbers may serve as an upper bound for a given set A, it should be noted that when the supremum of A exists it is unique (see exercise 4). We shall denote the supremum and infimum of A by $\sup(A)$ and $\inf(A)$, respectively.

In the field **R**, $\sup([0, 3)) = 3$ and $\inf([0, 3)) = 0$. If $A = \{2^{-k} : k \in \mathbf{N}\}$, then $\sup(A) = \frac{1}{2}$ and $\inf(A) = 0$. For $B = \{x \in \mathbf{R} : x^2 < 2\}$, $\sup(B) = \sqrt{2}$ and $\inf(B) = -\sqrt{2}$. Also, $\inf(\mathbf{N}) = 1$ and $\sup(\mathbf{N})$ does not exist.

In the field of rationals, if $A = \{2^{-k} : k \in \mathbf{N}\}$, then $\sup(A) = \frac{1}{2}$ and $\inf(A) = 0$. For $B = \{x \in \mathbf{Q} : x^2 < 2\}$, even though B is bounded in **Q**, B has no supremum or infimum in **Q**.

The following theorem provides a characterization of the supremum of a set.

Theorem 7.1. Let A be a subset of an ordered field F. Then $s = \sup(A)$ iff

(1) for all $\epsilon > 0$, if $x \in A$ then $x < s + \epsilon$.
(2) for all $\epsilon > 0$, there exists $y \in A$ such that $y > s - \epsilon$.

Proof. First suppose $s = \sup(A)$. Let $\epsilon > 0$ be given. Then $x \leq s < s + \epsilon$ for all $x \in A$, which establishes property (1).

To verify (2), suppose there were no y such that $y > s - \epsilon$. Then $s - \epsilon$ is an upper bound for A less than the least upper bound of A, a contradiction.

Suppose now that s is a number that satisfies conditions (1) and (2). To show that $s = \sup(A)$, we must first show that s is an upper bound for A. Suppose there is $y \in A$ such that $y > s$. If we let $\epsilon = (y - s)/2$, then $y > s + \epsilon$, which violates condition (1). Hence $y \leq s$ for all $y \in A$ and s is an upper bound.

To show s is the least of all upper bounds, suppose that there is another upper bound t such that $t < s$. If we let $\epsilon = s - t$ then, by condition (2), there is a number $y \in A$ such that $y > s - \epsilon$. Thus $y > s - \epsilon = s - (s - t) = t$. Hence $y > t$ and t is not an upper bound. Therefore, s is indeed the least upper bound of A. ■

We have seen that the field of rational numbers contains bounded subsets with no supremums in $\mathbf{Q}$ and therefore $\mathbf{Q}$ is not complete in this sense. On the other hand every bounded subset of the reals does have a supremum; a proof of this requires considerable preliminary study of exactly what is a real number and will be omitted. The distinction we seek between $\mathbf{R}$ and $\mathbf{Q}$ is given in the next definition.

Definition. An ordered field F is **complete** iff

(10) every nonempty subset of F that has an upper bound in F has a supremum in F.

The rationals are not complete. We state without proof that the reals are a complete ordered field. In fact, with much work we could show that if F and F' are any two complete ordered fields then they are essentially the same (isomorphic). This means there is a one-to-one correspondence from F to F' that preserves both operations, the order relations, and all supremums. Thus you should think of the terms "real number system" and "complete ordered field" as synonymous.

Exercises 7.1

1. Find the supremum and infimum in the real numbers, if they exist, of each of the following:

★ (a) $\left\{\frac{1}{n}: n \in \mathbf{N}\right\}$

(b) $\left\{\frac{n+1}{n}: n \in \mathbf{N}\right\}$

★ (c) $\{2^x: x \in \mathbf{Z}\}$

(d) $\left\{(-1)^n\left(1 + \frac{1}{n}\right): n \in \mathbf{N}\right\}$

★ (e) $\left\{\frac{n}{n+2}: n \in \mathbf{N}\right\}$

(f) $\{x \in \mathbf{Q}: x^2 < 10\}$

★ (g) $\{x: -1 \le x \le 1\} \cup \{5\}$

(h) $\{x: -1 \le x \le 1\} - \{0\}$

(i) $\left\{\frac{x}{2^y}: x, y \in \mathbf{N}\right\}$

(j) $\{x: |x| > 2\}$

2. Prove that if x is an upper bound for $A \subseteq \mathbf{R}$ and $x < y$, then y is an upper bound for A.
3. Prove that if u is an upper bound for $A \subseteq \mathbf{R}$ and $u \in A$ then $u = \sup(A)$.
4. Let $A \subseteq \mathbf{R}$.

★ (a) Prove that $\sup(A)$ is unique. That is, prove that if x and y are both least upper bounds for A, then $x = y$.

(b) Prove that $\inf(A)$ is unique.

5. Let $A \subseteq B \subseteq \mathbf{R}$.
   (a) Prove that if $\sup(A)$ and $\sup(B)$ both exist, then $\sup(A) \leq \sup(B)$.
   (b) Prove that if $\inf(A)$ and $\inf(B)$ both exist, then $\inf(A) \geq \inf(B)$.
6. Formulate and prove a characterization of greatest lower bounds similar to that in Theorem 7.1 for least upper bounds.
7. If possible, give an example of
   (a) a set $A \subseteq \mathbf{R}$ such that $\sup(A) = 4$ and $4 \notin A$.
   (b) a set $A \subseteq \mathbf{Q}$ such that $\sup(A) = 4$ and $4 \notin A$.
   (c) a set $A \subseteq \mathbf{N}$ such that $\sup(A) = 4$ and $4 \notin A$.
   (d) a set $A \subseteq \mathbf{N}$ such that $\sup(A) > 4$ and $4 \notin A$.
8. Give an example of a set of rational numbers that has a rational lower bound but no rational greatest lower bound.
9. Let $A \subseteq \mathbf{R}$.
   ★ (a) Prove that if $\sup(A)$ exists, then $\sup(A) = \inf\{u: u \text{ is an upper bound of } A\}$.
   (b) Prove that if $\sup(A)$ exists, then $\inf(A) = \sup\{\ell: \ell \text{ is a lower bound of } A\}$.
10. Prove that if $B \subseteq A \subseteq \mathbf{R}$ and A is bounded then B is bounded.
11. For $A \subseteq \mathbf{R}$, let $A^- = \{x \mid -x \in A\}$. Prove that if $\sup(A)$ exists, then $\inf(A^-)$ exists and $\inf(A^-) = -\sup(A)$.
12. Let A and B be subsets of $\mathbf{R}$.
   ★ (a) Prove that if $\sup(A)$ and $\sup(B)$ exist, then $\sup(A \cup B)$ exists and $\sup(A \cup B) = \max\{\sup(A), \sup(B)\}$.
   (b) State and prove a similar result for $\inf(A \cup B)$.
   (c) State and prove a relationship between $\sup(A)$, $\sup(B)$, and $\sup(A \cap B)$.
   (d) State and prove a relationship between $\inf(A)$, $\inf(B)$, and $\inf(A \cap B)$.
13. **Proofs to Grade.**
   ★ (a) **Claim.** Let $A \subseteq \mathbf{R}$. If $i = \inf(A)$ and $\epsilon > 0$, then there is $y \in A$ such that $y < i + \epsilon$.
   "Proof." Let $y = i + \epsilon/2$. Then $i < y$ so $y \in A$. By construction, $y < i + \epsilon$. ■
   (b) **Claim.** Let $(F, +, \cdot)$ be an ordered field. Then $0 < 1$.
   "Proof." By property (4), $0 \neq 1$. Therefore, by property (7), $0 < 1$ or $1 < 0$. If $1 < 0$, then $-1 > 0$. Thus $(-1)(-1) > 0(-1)$. Therefore, $1 > 0$, a contradiction. Thus $0 < 1$. ■
   (c) **Claim.** If $f: \mathbf{R} \to \mathbf{R}$ and A is a bounded subset of $\mathbf{R}$, then $f(A)$ is bounded.
   "Proof." Let m be an upper bound for A. Then $a \leq m$ for all $a \in A$. Therefore, $f(a) \leq f(m)$ for all $a \in A$. Thus $f(m)$ is an upper bound for $f(A)$. ■

SECTION 7.2. THE HEINE-BOREL THEOREM

In the next three sections we shall limit our discussion to the field of real numbers. It is not necessary to do so, but this provides a familiar model of an ordered field in which to work.

> **Definition.** For a real number a, if δ is a positive real, the **δ-neighborhood of a** is the set $\mathcal{N}(a, \delta) = \{x \in \mathbf{R}: |x - a| < \delta\}$.

Because $|x - a| < \delta$ is equivalent to $a - \delta < x < a + \delta$, the δ-neighborhood of a is the open interval $(a - \delta, a + \delta)$ centered at a with radius δ. Thus, for example, $\mathcal{N}(3, .5) = (2.5, 3.5)$ and $\mathcal{N}(1, .01) = (.99, 1.01)$.

> **Definition.** For a set $A \subseteq \mathbf{R}$, a point x is an **interior point of A** iff there exists $\delta > 0$ such that $\mathcal{N}(x, \delta) \subseteq A$. The set A is **open** in $\mathbf{R}$ iff every point of A is an interior point of A. The set A is **closed** in $\mathbf{R}$ iff its complement $\tilde{A}$ is open in $\mathbf{R}$.

For the interval $[2, 5)$, 3 is an interior point since $\mathcal{N}(3, .5) \subseteq [2, 5)$. Also, 4.98 is an interior point because $\mathcal{N}(4.98, .01) \subseteq [2, 5)$. See figure 7.1. In fact, every point in $(2, 5)$ is an interior point of $[2, 5)$. The point 2 is not interior to $[2, 5)$ since every δ-neighborhood of 2 contains points that are less than 2 and hence not in $[2, 5)$. Thus the set $[2, 5)$ is not open. Should we conclude that $[2, 5)$ is closed? Its complement $\widetilde{[2, 5)} = (-\infty, 2) \cup [5, \infty)$ contains 5 but not as an interior point. Thus $\widetilde{[2, 5)}$ is not open, so $[2, 5)$ is not closed.

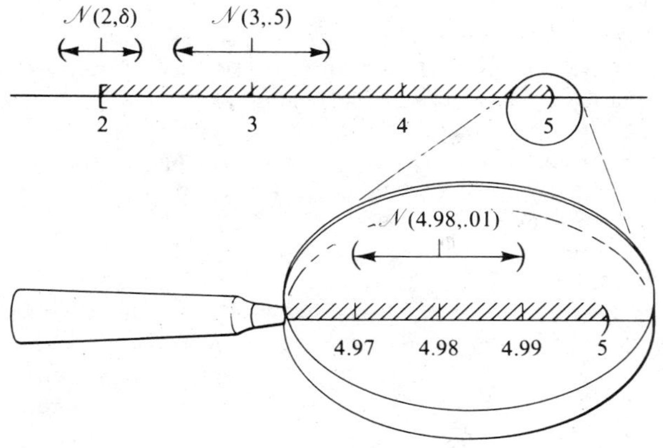

Figure 7.1

With practice you should come to recognize open subsets of $\mathbf{R}$. The next two theorems will help.

Theorem 7.2. Every open interval is an open set.

> ***Proof.*** Let (a, b) be an open interval. To show (a, b) is open, we let $x \in (a, b)$ and show x is an interior point of (a, b). ⟨ *That is, show $\mathcal{N}(x, \delta) \subseteq (a, b)$ for some $\delta > 0$.* ⟩ We choose $\delta = \min\{x - a, b - x\}$. ⟨ *The minimum is the largest possible δ we can use. See figure 7.2.* ⟩ Then $\delta > 0$. To show that $\mathcal{N}(x, \delta) \subseteq (a, b)$, let $y \in \mathcal{N}(x, \delta)$. Then $a \le x - \delta < y < x + \delta \le b$; so $y \in (a, b)$. ■

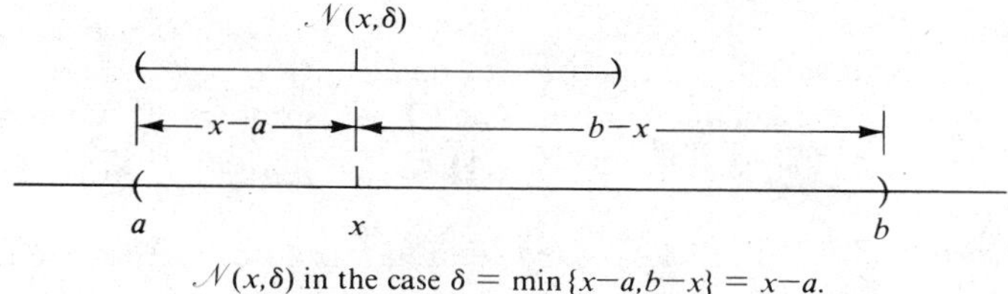

$\mathcal{N}(x,\delta)$ in the case $\delta = \min\{x-a, b-x\} = x-a$.

Figure 7.2

Theorem 7.3.

(a) Both $\varnothing$ and $\mathbf{R}$ are open sets.
(b) If $\mathscr{A}$ is a collection of open sets then $\bigcup_{A \in \mathscr{A}} A$ is open.
(c) If $\mathscr{A}$ is a finite collection of open sets then $\bigcap_{A \in \mathscr{A}} A$ is open.

Proof.

(a) Since $\mathcal{N}(x, 1) \subseteq \mathbf{R}$ for all $x \in \mathbf{R}$, $\mathbf{R}$ is open. Also $\varnothing$ is open, vacuously.

(b) Let $x \in \bigcup_{A \in \mathscr{A}} A$. Then there exists $B \in \mathscr{A}$ such that $x \in B$. Since B is in the collection $\mathscr{A}$, B is open and thus x is an interior point of B. Therefore, there exists $\delta > 0$ such that $\mathcal{N}(x, \delta) \subseteq B$. Since $B \subseteq \bigcup_{A \in \mathscr{A}} A$, $\mathcal{N}(x, \delta) \subseteq \bigcup_{A \in \mathscr{A}} A$. Therefore, x is an interior point of $\bigcup_{A \in \mathscr{A}} A$, which proves $\bigcup_{A \in \mathscr{A}} A$ is open.

(c) Let $x \in \bigcap_{A \in \mathscr{A}} A$. Then $x \in A$ for all $A \in \mathscr{A}$, and so for each open set $A \in \mathscr{A}$ there corresponds $\delta_A > 0$ such that $\mathcal{N}(x, \delta_A) \subseteq A$. By letting $\delta = \min\{\delta_A : A \in \mathscr{A}\}$ we have $\delta > 0$ and $\mathcal{N}(x, \delta) \subseteq A$ for all $A \in \mathscr{A}$. Thus $\mathcal{N}(x, \delta) \subseteq \bigcap_{A \in \mathscr{A}} A$, which proves $\bigcap_{A \in \mathscr{A}} A$ is open.

⟨*Where in this proof did we use the hypothesis that the collection $\mathscr{A}$ is finite?*⟩ ■

In the exercises you are invited to state and prove a theorem similar to Theorem 7.3 for closed sets. (See exercise 8.) Some examples of closed sets are $\mathbf{R}$ and $\varnothing$, any closed interval, and every finite subset of $\mathbf{R}$. (See exercise 5.) The set $\mathbf{Z}$ is closed in $\mathbf{R}$ because its complement, $\mathbf{R} - \mathbf{Z} = \bigcup_{a \in \mathbf{Z}} (a, a+1)$, is a union of open sets and hence is open.

The remainder of this section will concentrate on closed and bounded sets.

Theorem 7.4. If A is a nonempty closed and bounded subset of $\mathbf{R}$ then $\sup(A) \in A$ and $\inf(A) \in A$.

Proof. Denote $\sup(A)$ by s and suppose $s \notin A$. Then $s \in \bar{A}$, which is open, since A is closed. Thus $\mathcal{N}(s, \delta) \subseteq \bar{A}$ for some positive δ. This implies $s - \delta$ is an upper bound for A, since the interval $(s - \delta, s + \delta)$ is a subset of $\overline{A}$. ⟨*Notice that no element of A is greater than or equal to* $s + \delta$, *because s is an upper bound for A.*⟩ This contradicts Theorem 7.1. Therefore, $s \in A$. The proof that $\inf(A) \in A$ is similar. ■

Theorem 7.5. If A is a closed set and x is a point for which $A \cap \mathcal{N}(x, \delta) \neq \emptyset$ for all $\delta > 0$, then $x \in A$.

Proof. Exercise 9. ■

Definition. A collection $\mathcal{A}$ of sets is a **cover** for a set C iff $C \subseteq \bigcup_{A \in \mathcal{A}} A$. A **subcover of** $\mathcal{A}$ for C is a subcollection $\mathcal{B}$ of $\mathcal{A}$ that is also a cover for C.

If we let $H_a = (-\infty, a)$ for each $a \in \mathbf{R}$, then $\mathcal{H} = \{H_a : a \in \mathbf{R}\}$ is a cover for the reals, since $\bigcup_{a \in \mathbf{R}} H_a = \mathbf{R}$. The collection $\mathcal{W} = \{H_a : a \in \mathbf{Z}\}$ is a subcover of $\mathcal{H}$; and $\mathcal{L} = \{H_a : a \in \mathbf{N}\}$ is a subcover of $\mathcal{W}$ and of $\mathcal{H}$. We note in passing that there is no finite subset of $\mathcal{H}$ that covers $\mathbf{R}$.

Let $A_n = (n - \frac{1}{n}, n + \frac{1}{n})$ for each $n \in \mathbf{N}$. The collection $\mathcal{A} = \{A_n : n \in \mathbf{N}\}$ is a cover for $\mathbf{N}$ that has no subcover other than itself. See figure 7.3.

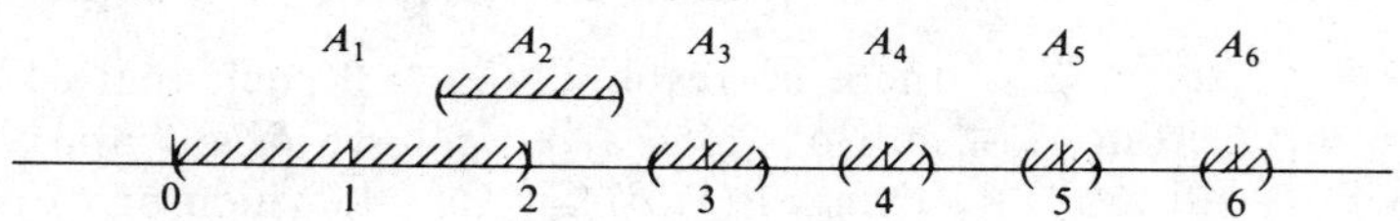

Figure 7.3

The collection $\mathcal{C} = \{(1/x, 1) : x \in (1, \infty)\}$ is a cover of the interval $(0, 1)$ by open sets. Then $\mathcal{D} = \{(1/x, 1) : x \in \mathbf{N} - \{1\}\}$ is a subcover of $\mathcal{C}$. The cover $\mathcal{C}$ of $(0, 1)$ also has no finite subcover. It also happens that $\mathcal{C}$ is a cover of the set $[.2, 1)$. Among the many finite subcovers of $\mathcal{C}$ for $[.2, 1)$ is $\mathcal{E} = \{(\frac{1}{2}, 1), (\frac{1}{11}, 1)\}$.

Definition. A subset A of $\mathbf{R}$ is **compact** iff every cover of A by open sets has a finite subcover.

The concept of compactness may be difficult to imagine initially. You should think of set A being compact as meaning that any time $A \subseteq \bigcup_{\alpha \in \Delta} O_\alpha$, a union of open sets, then there are a finite number of O_{α_i}, $i = 1, \ldots, n$ such that $A \subseteq \bigcup_{i=1}^{n} O_{\alpha_i}$.

We have seen $\mathscr{H} = \{(-\infty, a): a \in \mathbf{R}\}$ is an open cover of $\mathbf{R}$ with no finite subcover. Thus $\mathbf{R}$ is not compact. Likewise, (0, 1) is not compact, since $\mathscr{C} = \{(1/x, 1): x \in (1, \infty)\}$ has no finite subcover. What kind of sets are compact? Certainly, any finite set F is compact; if $F \subseteq \bigcup_{\alpha \in \Delta} O_\alpha$ where each O_α is open, simply select O_{α_x} such that $x \in O_{\alpha_x}$ for each $x \in F$; then $\{O_{\alpha_x}: x \in F\}$ is a finite subcover.

We close this section with an elegant characterization of the compact subsets of $\mathbf{R}$. The next theorem can be traced to the works of Edward Heine (1821–1881) and Emile Borel (1871–1938). Watch how heavily the proof depends upon the completeness of $\mathbf{R}$.

Theorem 7.6. (The Heine-Borel Theorem.) A subset A of $\mathbf{R}$ is compact iff A is closed and bounded.

Proof.

(i) Suppose A is compact. We first show A is bounded. We note that $\mathbf{R} = \bigcup_{n \in \mathbf{N}} (-n, n)$; thus $A \subseteq \bigcup_{n \in \mathbf{N}} (-n, n)$. Therefore, $\mathscr{H} = \{(-n, n): n \in \mathbf{N}\}$ is an open cover of A. By compactness, $\mathscr{H}$ has a finite subcover $\mathscr{H}' = \{(-n, n) \mid n \in \{n_1, n_2, \ldots, n_k\}\}$. If we choose $N = \max\{n_1, n_2, \ldots, n_k\}$, then $A \subseteq (-N, N)$. Therefore A is bounded.

We next show A is closed by proving $\bar{A}$ is open. Let $y \in \bar{A}$. ⟨*We must show y is an interior point of* $\bar{A}$.⟩ For each $x \in A$, $x \neq y$ and thus $\delta_x = \frac{1}{2}|x - y|$ is a positive number. The collection $\{\mathscr{N}(x, \delta_x): x \in A\}$ is a family of open sets that covers A. Hence by compactness $A \subseteq \mathscr{N}(x_1, \delta_{x_1}) \cup \mathscr{N}(x_2, \delta_{x_2}) \cup \ldots \cup \mathscr{N}(x_k, \delta_{x_k})$ for some $x_1, x_2, \ldots, x_k \in A$. By choosing $\delta = \min\{\delta_{x_1}, \delta_{x_2}, \ldots, \delta_{x_k}\}$, we have $\mathscr{N}(y, \delta) \subseteq \bar{A}$. ⟨*If* $z \in A$, *then* $|z - x_i| < \delta_{x_i}$ *for some i. Thus if* $z \in \mathscr{N}(y, \delta)$, *then* $|z - y| < \delta \leq \delta_{x_i}$ *and* $|x_i - y| \leq |x_i - z| + |z - y| < 2\delta_{x_i} = |x_i - y|$.⟩ Thus $\bar{A}$ is open. Hence A is closed. See figure 7.4.

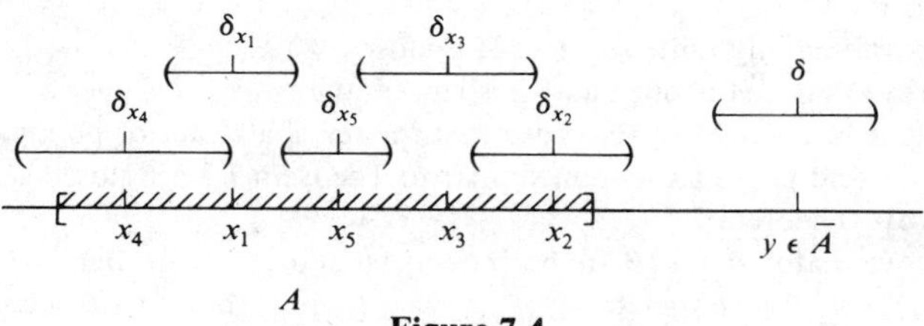

Figure 7.4

(ii) Conversely, suppose A is a closed and bounded set and $\mathscr{C}$ is an open cover for A. ⟨*We show* $\mathscr{C}$ *has a finite subcover.*⟩ For each $x \in \mathbf{R}$, let

$A_x = \{a \in A : a \leq x\}$. Also let $D = \{x \in \mathbf{R} : A_x$ is included in a union of finitely many sets from $\mathscr{C}\}$.

Since A is bounded, $\inf(A)$ exists. Thus if $x < \inf(A)$, $A_x = \varnothing$ and it follows that $x \in D$. Therefore, $(-\infty, \inf(A)) \subseteq D$ and so D is nonempty.

We claim D has no upper bound. If D is bounded above, then $x_0 = \sup(D)$ exists ⟨*by the completeness of* **R**.⟩ Let $\delta > 0$ and choose $t \in D$ such that $x_0 - \delta < t < x_0$ ⟨*applying Theorem 7.1*⟩. If $A \cap \mathscr{N}(x_0, \delta) = \varnothing$, then $A_t = \{a \in A : a \leq t\} = \{a \in A : a \leq x_0 + \frac{\delta}{2}\} = A_{x_0 + \frac{\delta}{2}}$. Since $t \in D$, $x_0 + \frac{\delta}{2} \in D$, which is a contradiction to $x_0 = \sup(D)$. Therefore, for all $\delta > 0$ we have $A \cap \mathscr{N}(x_0, \delta) \neq \varnothing$. Since A is closed, $x_0 \in A$. Let C^* be an element of $\mathscr{C}$ such that $x_0 \in C^*$. Since C^* is open there exists $\epsilon > 0$ such that $\mathscr{N}(x_0, \epsilon) \subseteq C^*$. Choose $x_1 \in D$ such that $x_0 - \epsilon < x_1 \leq x_0$. Since $x_1 \in D$ there are open sets $C_1, C_2, \ldots, C_n$ in $\mathscr{C}$ such that $A_{x_1} \subseteq C_1 \cup C_2 \cup \ldots \cup C_n$. Now let $x_2 = x_0 + \frac{\epsilon}{2}$. Then $x_2 \in C^*$ and
$A_{x_2} \subseteq C_1 \cup C_2 \cup \ldots \cup C_n \cup C^*$. Thus $x_2 \in D$, a contradiction, since $x_2 > x_0$ and $x_0 = \sup(D)$. We conclude that D has no upper bound.

Finally since D has no upper bound, choose $x \in D$ such that $x > \sup(A)$ ⟨*sup(A) exists because* **R** *is complete.*⟩ Thus $A_x = A$ and since $x \in D$, A is included in a union of finitely many sets from $\mathscr{C}$. Therefore A is compact. ■

Exercises 7.2

1. Classify each of the following subsets of **R** as open, closed, both open and closed, or neither open nor closed.
   (a) $(-\infty, -3)$
   ★ (b) $\mathscr{N}(a, \delta) - \{a\}$, $a \in \mathbf{R}$.
   (c) $(5, 8) \cup \{9\}$
   (d) **Q**
   ★ (e) **R** − **N**
   (f) $\{x : |x - 5| = 7\}$
   (g) $\{x : |x - 5| > 7\}$
   (h) $\{x : |x - 5| \neq 7\}$
   ★ (i) $\{x : |x - 5| \leq 7\}$
   (j) $\{x : 0 < |x - 5| \leq 7\}$
2. ☆ Prove that if A and B are closed subsets of **R**, then $A \cup B$ is closed.
3. Give an example of a family of sets $\mathscr{A}$ such that each $A \in \mathscr{A}$ is closed but $\bigcup_{A \in \mathscr{A}} A$ is not closed.
4. Give an example of a family of sets $\mathscr{A}$ such that each $A \in \mathscr{A}$ is open but $\bigcap_{A \in \mathscr{A}} A$ is not open.
5. ★ Prove that any finite set $A \subseteq \mathbf{R}$ is closed.
6. Prove that if A is open and B is closed, then $A - B$ is open.
7. Let A be a subset of **R**. Prove that the set of all interior points of A is an open set.
8. State and prove a theorem similar to Theorem 7.3 for closed sets.
9. Prove Theorem 7.5.
10. Prove that if A and B are both compact subsets of **R**, then
    ☆ (a) $A \cup B$ is compact.
    (b) $A \cap B$ is compact.
11. Give an example of an open bounded subset of **R** and an open cover of that set which has no finite subcover.
12. Give an example of a closed subset of **R** and an open cover of that set which has no finite subcover.

13. Which of the following subsets of $\mathbf{R}$ are compact?
★ (a) $\mathbf{Z}$
(b) $[0, 10] \cup [20, 30]$
(c) $[\pi, \sqrt{10}]$
★ (d) $\mathbf{R} - A$, where A is any finite set
(e) $\{1, 2, 3, 4, 9, 12, 18\}$
(f) $\{0\} \cup \left\{\frac{1}{n} : n \in \mathbf{N}\right\}$
★ (g) $(-3, 5]$
(h) $[0, 1] \cap \mathbf{Q}$

☆ 14. Give two different proofs that any finite set is compact.

15. Let $S = (0, 1]$ and let

$$\mathscr{C} = \left\{\left(\frac{n+2}{2^n}, 2^{1/n}\right) : n \in \mathbf{N}\right\}$$

(a) Prove that $\mathscr{C}$ is an open cover of S.
(b) Is there a finite subcover of $\mathscr{C}$ which covers S?
(c) What does the Heine-Borel Theorem say about S?

16. Use the Heine-Borel Theorem to prove that if $\{A_\alpha : \alpha \in \Delta\}$ is a collection of compact sets, then $\bigcap_{\alpha \in \Delta} A_\alpha$ is compact.

17. Give an example of a collection $\{A_\alpha : \alpha \in \Delta\}$ of compact sets such that $\bigcup_{\alpha \in \Delta} A_\alpha$ is not compact.

18. **Proofs to Grade.**
(a) **Claim.** If A and B are compact then $A \cup B$ is compact.
"Proof." Suppose A and B are compact. Then for any open cover $\{O_\alpha : \alpha \in \Delta\}$ of A there exists a finite subcover $O_{\alpha_1}, O_{\alpha_2}, \ldots, O_{\alpha_n}$, and for any open cover $\{U_\beta : \beta \in \Gamma\}$ of B there exists a finite subcover $U_{\beta_1}, U_{\beta_2}, \ldots, U_{\beta_m}$. Thus $A \subseteq O_{\alpha_1} \cup O_{\alpha_2} \cup \ldots \cup O_{\alpha_n}$ and $B \subseteq U_{\beta_1} \cup U_{\beta_2} \cup \ldots \cup U_{\beta_m}$. Therefore, $A \cup B \subseteq O_{\alpha_1} \cup O_{\alpha_2} \cup \ldots \cup O_{\alpha_n} \cup U_{\beta_1} \cup U_{\beta_2} \cup \ldots \cup U_{\beta_m}$, a union of a finite number of open sets. Thus $A \cup B$ is compact. ■

★ (b) **Claim.** If A is compact, $B \subseteq A$ and B is closed, then B is compact.
"Proof." Let $\{O_\alpha : \alpha \in \Delta\}$ be an open cover of B. If $\{O_\alpha : \alpha \in \Delta\}$ is an open cover of A, then there is a finite subcover of $\{O_\alpha : \alpha \in \Delta\}$ which covers A and hence covers B. If $\{O_\alpha : \alpha \in \Delta\}$ is not an open cover of A, add one more open set $O^* = A - B$ to the collection to obtain an open cover of A. This open cover of A has a finite subcover which is a cover of B. In either case B is covered by a finite number of open sets. Therefore, B is compact. ■

(c) **Claim.** If A is compact, $B \subseteq A$, and B is closed, then B is compact.
"Proof." B is closed by assumption. Since A is compact, A is bounded. Since $B \subseteq A$, B is also bounded. Thus B is closed and bounded. Therefore B is compact. ■

SECTION 7.3. THE BOLZANO-WEIERSTRASS THEOREM

The set $A = \{(-1)^{n+1}/n : n \in \mathbf{N}\}$ is bounded above by 1 and below by $-1/2$. Since A is infinite it would seem that its points must pile up or "accumulate" around some number or numbers between $-1/2$ and 1. The number 0 is a point about which points of A accumulate. (See figure 7.5.)

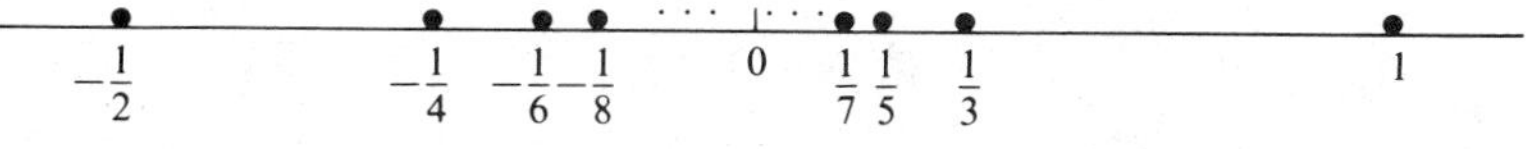

Figure 7.5

Definition. The number x is an **accumulation point for the set A** iff for all $\delta > 0$, $\mathcal{N}(x, \delta)$ contains a point of A distinct from x.

An accumulation point might or might not belong to the set in question. The number 0 is the only accumulation point of $\{(-1)^{n+1}/n : n \in \mathbf{N}\}$, although 0 is not in the set. On the other hand, the closed interval $[0, 1]$ has every one of its points as an accumulation point and no others. The set $\{(-1)^n(n+1)/n) : n \in \mathbf{N}\}$ has exactly two accumulation points, -1 and 1.

It is important to remember that every δ-neighborhood of an accumulation point x of a set A contains points of A **other than x.** This means that no finite set A can have an accumulation point since $\mathcal{N}(x, \delta) \cap A = \{x\}$ for all $x \in A$ if we choose $\delta = \min\{|x - y| : x, y \in A, x \neq y\}$.

The set of accumulation points of A is called the **derived set of A,** denoted A'. As examples, $(0, 1)' = [0, 1]$ and $\mathbf{N}' = \emptyset$. If $B = \{1.4, 1.41, 1.414, 1.4142, 1.41421, \ldots\}$ is a set of successive decimal approximations to $\sqrt{2}$, then $B' = \{\sqrt{2}\}$. The following theorem relates derived sets and closed sets.

Theorem 7.7. A set A is closed iff $A' \subseteq A$.

Proof. Suppose A is closed and $x \in A'$. We prove $x \in A$ by contradiction. If $x \notin A$, then $x \in \tilde{A}$, an open set. Thus, $\mathcal{N}(x, \delta) \subseteq \tilde{A}$ for some positive δ. But then $\mathcal{N}(x, \delta)$ can contain no points of A except perhaps for x. Thus $x \notin A'$, a contradiction. We conclude $x \in A$.

Suppose now $A' \subseteq A$. To show A is closed we show $\tilde{A}$ is open by contradiction. If $\tilde{A}$ is not open there is at least one $x \in \tilde{A}$ that is not an interior point of $\tilde{A}$. Therefore, no δ-neighborhood of x is a subset of $\tilde{A}$; that is, each δ-neighborhood of x contains a point of A other than x, since $x \in \tilde{A}$. Thus $x \in A'$. But $A' \subseteq A$, so $x \in A$, a contradiction. ■

At the beginning of this section we inferred that the bounded infinite set $A = \{(-1)^{n+1}/n : n \in \mathbf{N}\}$ must have at least one accumulation point. We are now in a position to prove this is so with the second classical result of this chapter, the Bolzano-Weierstrass Theorem.

Bernard Bolzano (1781–1849) and Karl Weierstrass (1815–1897) were leaders in developing highly rigorous standards of mathematical analysis. Weierstrass has been called the father of modern analysis. The proof that follows rests upon the Heine-Borel Theorem.

Theorem 7.8. (The Bolzano-Weierstrass Theorem.) Every bounded infinite subset of real numbers has an accumulation point.

Proof. Suppose the set A is bounded, infinite, but has no accumulation points. Then A is closed by Theorem 7.7. $\langle$*$A' = \emptyset$, hence $A' \subseteq A$*$\rangle$. Thus, by the Heine-Borel Theorem, A is compact.

Since A has no accumulation points, for each $x \in A$, there exists $\delta_x > 0$ such that $\mathcal{N}(x, \delta_x) \cap A = \{x\}$. Thus, for $x \neq y$ in A, $\mathcal{N}(x, \delta_x) \neq \mathcal{N}(y, \delta_y)$. But then the family $\mathcal{A} = \{\mathcal{N}(x, \delta_x): x \in A\}$ is an infinite collection of open sets that covers A with no subcover other than itself; hence, it has no finite subcover. This contradicts the fact that A is compact. Therefore A must have an accumulation point. ∎

Exercises 7.3

1. Find the derived set of each of the following sets.
   ★ (a) $\left\{\frac{n+1}{2n}: n \in \mathbf{N}\right\}$
   (b) $\{2^n: n \in \mathbf{N}\}$
   ★ (c) $\{6n: n \in \mathbf{N}\}$
   (d) $\{m/2^n: n, m \in \mathbf{N}\}$
   (e) $(0, 1]$
   (f) $(3, 7) \cup \{4, 6, 8\}$
   ★ (g) $\{1 + \frac{(-1)^n n}{n+1}: n \in \mathbf{N}\}$
   (h) $\mathbf{Z}$
   (i) $\mathbf{Q}$
   (j) $\left\{\frac{1 + n^2(1 + (-1)^n)}{n}: n \in \mathbf{N}\right\}$
2. Find an example of an infinite subset of $\mathbf{R}$ that has
   (a) no accumulation points.
   (b) exactly one accumulation point.
   (c) exactly two accumulation points.
   (d) denumerably many accumulation points.
   (e) an uncountable number of accumulation points.
3. ☆ Prove that if $A \subseteq \mathbf{R}$, $z = \sup(A)$ and $z \notin A$, then z is an accumulation point of A.
4. (a) Prove that if $A \subseteq B \subseteq \mathbf{R}$, then $A' \subseteq B'$.
   (b) Is the converse of part (a) true? Explain.
5. Prove that $(A \cup B)' = A' \cup B'$. (The operation of finding the derived set preserves unions.)
6. Prove that $(A \cap B)' \subseteq A' \cap B'$. Show by example that set equality need not hold.
7. Prove that if B is closed and $A \subseteq B$, then $A' \subseteq B$.
8. Define the **closure** of a set A to be $c(A) = A \cup A'$.
   (a) Prove that $c(A)$ is a closed set.
   (b) Prove that $A \subseteq c(A)$.
   ★ (c) Prove that $c(A)$ is the smallest closed set containing A. That is, if B is closed and $A \subseteq B$, show that $c(A) \subseteq B$.
   (d) Prove that the derived set of the closure of A is the same as the derived set of A.
9. Which of the following have at least one accumulation point?
   ★ (a) An infinite subset of $\mathbf{N}$.
   (b) An infinite subset of $(-10, 10)$.
   ★ (c) An infinite subset of $[0, 100]$.
   (d) An infinite subset of $\mathbf{Q}$.
10. Prove that if a bounded subset of the real numbers has no accumulation points, the set is finite.
11. Let x be an accumulation point of the set A and let F be a finite set. Prove that x is an accumulation point of $A - F$.
12. ★ Let $S = (0, 1]$. Find $S' \cap (\bar{S})'$.
13. Prove that if $S \subseteq \mathbf{R}$ is open then every point of S is an accumulation point of S.
14. **Proofs to Grade.**
   ☆ (a) **Claim.** For $A, B \subseteq \mathbf{R}$, $(A \cup B)' = A' \cup B'$.
   "Proof."
      (i) Since $A \subseteq A \cup B$, $A' \subseteq (A \cup B)'$ by problem 4 (a). Likewise, $B' \subseteq (A \cup B)'$. Therefore $A' \cup B' \subseteq (A \cup B)'$.

(ii) To show $(A \cup B)' \subseteq A' \cup B'$, let $x \in (A \cup B)'$. Then, for all $\delta > 0$, $\mathcal{N}(x, \delta)$ contains a point of $A \cup B$ distinct from x. Restating this, we have, for all $\delta > 0$, that $\mathcal{N}(x, \delta)$ contains a point of A distinct from x or a point of B distinct from x. Thus, for all $\delta > 0$, $\mathcal{N}(x, \delta)$ contains a point of A distinct from x, or, for all $\delta > 0$, $\mathcal{N}(x, \delta)$ contains a point of B distinct from x. But this means $x \in A'$ or $x \in B'$. Therefore, $x \in A' \cup B'$. ■

(b) **Claim.** If A is a set with an accumulation point, $B \subseteq A$, and B is infinite, then B has an accumulation point.
"Proof." First, A is infinite because $B \subseteq A$ and B is infinite. Since A has an accumulation point, by the Bolzano-Weierstrass Theorem, A must be bounded. Since $B \subseteq A$, this means B is bounded. Hence by the Bolzano-Weierstrass Theorem again, B has an accumulation point. ■

★ (c) **Claim.** For $A, B \subseteq \mathbf{R}$, $(A - B)' \subseteq A' - B'$.
"Proof:"

$$\begin{aligned}(A - B)' &= (A \cap \tilde{B})' \quad \langle \textit{definition of } A - B \rangle \\ &\subseteq A' \cap (\tilde{B})' \quad \langle \textit{problem 6} \rangle \\ &\subseteq A' \cap \widetilde{(B')} \quad \langle \textit{since } (\tilde{B})' \subseteq \widetilde{(B')} \rangle \\ &= A' - B'.\end{aligned}$$

■

SECTION 7.4. THE BOUNDED MONOTONE SEQUENCE THEOREM

In this section we shall use the Bolzano-Weierstrass Theorem to prove an important result about sequences. The concept of a sequence should be familiar to you from calculus.

A **sequence** is a function x whose domain is $\mathbf{N}$. A sequence of real numbers, or **real sequence,** has codomain $\mathbf{R}$. Every sequence we consider will be a real sequence. For $n \in \mathbf{N}$, the image of n under the sequence x is written as x_n rather than the customary functional notation $x(n)$. We call x_n the ***n*th term of the sequence.**

Suppose x is the sequence given by $x_n = 2^n$. The first few terms are $x_1 = 2$, $x_2 = 4$, $x_3 = 8$, and so on. The sequence y, where $y_n = (-1)^n$, has the alternating terms -1 and 1.

A sequence x is **bounded** iff its range $\{x_n : n \in \mathbf{N}\}$ is a bounded subset of $\mathbf{R}$. Boundedness may be described in terms of absolute value.

Theorem 7.9. A sequence x of real numbers is bounded iff there exists a number B such that $|x_n| \leq B$ for all $n \in \mathbf{N}$.

Proof. Exercise 7. ■

The sequence given by $x_n = 2^n$ is bounded below but not bounded, whereas the sequence y defined by $y_n = (-1)^n$ is bounded by 1. The sequence whose nth term is $z_n = e^{-n}$ is bounded by e^{-1}.

The bounded sequence x given by $x_n = (n + 1)/n$ has for its first few terms, $2, \frac{3}{2}, \frac{4}{3}, \frac{5}{4}, \ldots$. Furthermore, for large values of n, the values of x_n are close to the number 1. This is the concept of the limit of x.

Definition. A sequence x **has the limit L,** or, **x converges to L,** iff, for every real $\epsilon > 0$, there exists a natural number N such that if $n > N$ then $|x_n - L| < \epsilon$. If x converges to L we write $\lim_{n\to\infty} x_n = L$ or $x_n \to L$. If no such L exists, the sequence **diverges.**

To prove a given sequence converges involves first deciding the value of the limit L and then verifying the definition of convergence. For example, suppose $x_n = 3n^2/(n^2 + 1)$. After calculating x_n for several large values of n, we make an educated guess that x converges to 3. Next we must show that the terms approach 3 by showing that, for every $\epsilon > 0$, there is a natural number N such that $n > N$ implies $|(3n^2/(n^2 + 1)) - 3| < \epsilon$. Since $|(3n^2/(n^2 + 1)) - 3| = |(-3/(n^2 + 1))| = 3/(n^2 + 1)$, we require an integer N such that $n > N$ implies $3/(n^2 + 1) < \epsilon$ or, equivalently, $n^2 + 1 < 3/\epsilon$. By selecting N to be any natural number greater than $3/\epsilon$, we have that $n > N$ implies (since $n^2 + 1 > n$) $n^2 + 1 > N > 3/\epsilon$. The above is scratchwork for the formal proof that follows.

Example. Prove that the sequence x given by $x_n = 3n^2/(n^2 + 1)$ converges.

Proof. We will show $\lim_{n\to\infty} 3n^2/(n^2 + 1) = 3$. Let $\epsilon > 0$. Let N be a natural number greater than $3/\epsilon$. Then since $n^2 + 1 > n$ for all n, if $n > N$ then $n > 3/\epsilon$, and thus $n^2 + 1 > 3/\epsilon$. Therefore, if $n > N$, then $3/(n^2 + 1) < \epsilon$ or, equivalently, $|3n^2/(n^2 + 1) - 3| < \epsilon$. Thus the definition is satisfied and $3n^2/(n^2 + 1) \to 3$. ■

Intuitively, a sequence x converges if the terms x_n get closer to a single number L for larger values of n. For this reason the sequence $y_n = (-1)^n$ must diverge. A proof of this must show the negation of the definition of a limit: if $\lim_{n\to\infty} y_n \neq L$, then there exists a real $\epsilon > 0$ such that for all $N \in \mathbf{N}$ there exists $n > N$ such that $|y_n - L| \geq \epsilon$.

Example. Prove that the sequence given by $y_n = (-1)^n$ has no limit.

Proof. Suppose $\lim_{n\to\infty} y_n = L$ for some number L. Let $\epsilon = 1$. We must show that for all $N \in \mathbf{N}$, there exists $n > N$ such that $|y_n - L| \geq 1$. Let $N \in \mathbf{N}$. If $L > 0$, let n be any odd integer greater than N, which means $y_n = -1$. For $L \leq 0$, let n be any even integer greater than N; here $y_n = 1$. In each of these cases n was selected so that $n > N$ and $|y_n - L| \geq 1$. Therefore, $\lim_{n\to\infty} y_n \neq L$. ■

Theorem 7.10. If a sequence x converges, then its limit is unique.

Proof. Suppose $x_n \to L$ and $x_n \to M$ and $L \neq M$. Let $\epsilon = \frac{1}{3}|L - M|$. ⟨ *The idea of the proof is to suppose there are two different limits and select ϵ so small that the terms cannot simultaneously be within ϵ of each limit.*⟩

Since $x_n \to L$ and $x_n \to M$, there are natural numbers N_1 and N_2 such that $n > N_1$ implies $|x_n - L| < \epsilon$ and $n > N_2$ implies $|x_n - M| < \epsilon$. Let $N = N_1 + N_2$. Then, since N is greater than both N_1 and N_2,

$$\begin{aligned} |L - M| &= |(L - x_n) + (x_n - M)| \\ &\leq |L - x_n| + |x_n - M| \\ &= |x_n - L| + |x_n - M| \\ &< \epsilon + \epsilon \\ &= 2\epsilon \\ &= \tfrac{2}{3}|L - M|. \end{aligned}$$

Thus the assumption $L \neq M$ leads to the contradiction $|L - M| < \frac{2}{3}|L - M|$. We conclude the limit of x is unique. ■

The sequence with nth term $x_n = 2^n$ diverges because the terms increase without bound and cannot stay "close" to any one real number. Turning this idea around, a convergent sequence must be bounded because all but the first few terms will be close to the limit.

Theorem 7.11. Every sequence that converges is bounded.

Proof. Suppose x is a sequence converging to a number L. For $\epsilon = 1$, there is a natural number N such that if $n > N$, then $|x_n - L| < 1$. Since $||x_n| - |L|| \leq |x_n - L|$, we have $|x_n| - |L| < 1$ for all $n > N$. Thus for all $n > N$, $|x_n| < |L| + 1$. ⟨ *All but the first N terms are bounded by* $|L| + 1$. ⟩ Let $B = \max\{|x_1|, |x_2|, \ldots, |x_N|, |L| + 1\}$. Then $|x_n| \leq B$ for all $n \in \mathbf{N}$ and x is bounded. ■

Definition. A sequence x is **increasing** iff for all $n, m \in \mathbf{N}$, $x_n \leq x_m$ whenever $n < m$. A **decreasing** sequence y requires $y_n \geq y_m$ for $n < m$. A **monotone** sequence is one that is either increasing or decreasing.

A constant sequence t, where $t_n = c$ for some fixed $c \in \mathbf{R}$ and for all n, is monotone since it is both increasing and decreasing. The sequence given by $x_n = 2^n$ is increasing while $z_n = e^{-n}$ is decreasing. The sequence y defined by $y_n = (-1)^n$ is not monotone.

The next theorem relates all the concepts of this section: a sequence that is both bounded and monotone must converge. Its proof uses the Bolzano-Weierstrass Theorem.

Theorem 7.12. (The Bounded Monotone Sequence Theorem.) For every bounded monotone sequence x, there is a real number L such that $x_n \to L$.

Proof. Assume x is a bounded and increasing sequence. The proof in the case where x is decreasing is similar.

If $\{x_n : n \in \mathbf{N}\}$ is finite, then let $L = \max\{x_n : n \in \mathbf{N}\}$. For some $N \in \mathbf{N}$, $x_N = L$ and, since x is increasing, $x_n = L$ for all $n > N$. Therefore, $x_n \to L$.

Suppose $\{x_n : n \in \mathbf{N}\}$ is infinite. Because x is bounded this set must have an accumulation point L, by the Bolzano-Weierstrass Theorem. We claim $x_n \leq L$ for all $n \in \mathbf{N}$. If there exists N such that $x_N > L$, then $x_n > L$ for all $n \geq N$. By exercise 11 of section 7.3, L is an accumulation point of $\{x_n : n \geq N\}$. But for $\delta = |x_N - L|$, $\mathcal{N}(L, \delta)$ contains no points of $\{x_n : n \geq N\}$. This is a contradiction. Thus $x_n \leq L$ for all n. The remainder of this proof will show $x_n \to L$.

Let $\epsilon > 0$. Since L is an accumulation point of $\{x_n : n \in \mathbf{N}\}$, there exists $M \in \mathbf{N}$ such that $x_M \in \mathcal{N}(L, \epsilon)$. Thus $L - \epsilon < x_M$, and so, for $n > M$,

$$L - \epsilon < x_M \leq x_n \leq L < L + \epsilon.$$

Therefore, for $n > M$, $|x_n - L| < \epsilon$. Thus $x_n \to L$. ■

Exercises 7.4

1. For each sequence x, prove that x converges or diverges.

☆ (a) $x_n = \dfrac{(-1)^n}{n}$
(b) $x_n = \dfrac{n+1}{n}$
☆ (c) $x_n = n^2$
(d) $x_n = \dfrac{(-1)^n n}{2n+1}$
(e) $x_n = \dfrac{\sin n}{n}$
☆ (f) $x_n = \sqrt{n+1} - \sqrt{n}$
(g) $x_n = \left(\dfrac{n}{2}\right)^n$
(h) $x_n = \dfrac{6}{2^n}$
(i) $x_n = \dfrac{5000}{n!}$
(j) $x_n = \sin\left(\dfrac{n\pi}{2}\right)$

2. Give an example of a bounded sequence that is not convergent.
3. Give an example of an increasing sequence that is not convergent.
4. Prove that if $x_n \to L$ and $y_n \to M$ and $r \in \mathbf{R}$, then

★ (a) $x_n + y_n \to L + M$
(b) $x_n - y_n \to L - M$
(c) $rx_n \to rL$
(d) $x_n y_n \to LM$

☆ 5. Prove that if $x_n \to 0$ and y is a bounded sequence, then $x_n y_n \to 0$.
6. ☆ (a) Prove that if $x_n \to L$, then $|x_n| \to |L|$.
   (b) Give an example of a sequence x such that $|x_n| \to |L|$ but x does not converge.
7. Prove Theorem 7.9.
8. Give a proof of The Bounded Monotone Sequence Theorem for the case in which the sequence x is assumed bounded and decreasing.
9. ☆ (a) Prove that if $x_n \to L$ and $L \neq 0$, then there is a number N such that if $n \geq N$, then $|x_n| > (|L|/2)$.
   (b) Prove that if $x_n \to L$, $x_n \neq 0$ for all n, $L \neq 0$, and if $y_n \to M$, then $(y_n/x_n) \to (M/L)$.
10. **Proofs to Grade.**
    (a) **Claim.** Every bounded decreasing sequence converges.
    "Proof." Let x be a bounded decreasing sequence. Then $y_n = -x_n$ defines a bounded increasing sequence. By the proof of Theorem 7.12, $y_n \to L$ for some L. Thus $x_n \to L$. ■

★ (b) **Claim.** If the sequence x converges and the sequence y diverges, then $x + y$ diverges.
"Proof." Suppose $x_n + y_n \to K$ for some real number K. Since $x_n \to L$ for some number L, $(x_n + y_n) - x_n \to K - L$; that is, $y_n \to K - L$. This is a contradiction. Thus $x_n + y_n$ diverges. ■

(c) **Claim.** If two sequences x and y both diverge, then $x + y$ diverges.
"Proof." Suppose $\lim_{n\to\infty} (x_n + y_n) = L$. Since x diverges, there exists $\epsilon_1 > 0$ such that for all $N \in \mathbf{N}$ there exist $n > N$ such that $|x_n - (L/2)| \geq \epsilon_1$. Since y diverges, there exists $\epsilon_2 > 0$ such that, for all $N \in \mathbf{N}$, there exists $n > N$ such that $|y_n - (L/2)| \geq \epsilon_2$. Let $\epsilon = \frac{1}{2}\min\{\epsilon_1, \epsilon_2\}$. Then, for all $N \in \mathbf{N}$, there exists $n > N$ such that

$$\begin{aligned}|(x_n + y_n) - L| &= |(x_n - (L/2)) + (y_n - (L/2))| \\ &\geq |x_n - (L/2)| + |y_n - (L/2)| \\ &\geq \epsilon_1 + \epsilon_2 \\ &\geq \tfrac{1}{2}\epsilon + \tfrac{1}{2}\epsilon \\ &= \epsilon.\end{aligned}$$

Therefore, $\lim_{n\to\infty} (x_n + y_n) \neq L$. ■

SECTION 7.5. EQUIVALENTS OF COMPLETENESS

In the first section of this chapter we stated without proof that the ordered field of real numbers is complete. Completeness was then used to prove the Heine-Borel Theorem in section 7.2. The Heine-Borel Theorem was the main result used to prove the Bolzano-Weierstrass Theorem in section 7.3, which in turn justified the Bounded Monotone Sequence Theorem of the last section. In this section we complete a circle of arguments by showing that if the ordered field of reals is assumed to have every bounded monotone sequence convergent, then the reals are complete. Thus completeness and the three properties described by the theorems are equivalent for $\mathbf{R}$. What has been omitted is a proof of the (correct) statement that $\mathbf{R}$ is complete. Furthermore, the theorems stated for the reals could be generalized to prove that the properties are all equivalent for any ordered field. Our proof of completeness from the Bounded Monotone Sequence Theorem will use the following two lemmas about convergent sequences. The proofs are exercises.

Lemma 7.13. If x and y are two sequences such that $\lim_{n\to\infty} y_n = s$ and $\lim_{n\to\infty} (x_n - y_n) = 0$, then $\lim_{n\to\infty} x_n = s$.

Lemma 7.14. If x is a sequence with $\lim_{n\to\infty} x_n = s$ and t is a real number such that $t < s$, then there exists $N \in \mathbf{N}$ such that $x_n > t$ for all $n \geq N$.

Theorem 7.15. Suppose the real number system has the property that every bounded monotone sequence must converge. Then $\mathbf{R}$ is complete.

Proof. Let A be a nonempty subset of $\mathbf{R}$ that is bounded above by a real number b. To prove completeness we must show $\sup(A)$ exists and is a real number. Since $A \neq \varnothing$, choose $a \in A$. If $(a + b)/2$ is an upper bound for A, let $x_1 = a$ and $y_1 = (a + b)/2$; if not let $x_1 = (a + b)/2$ and $y_1 = b$. In either case, $y_1 - x_1 = (b - a)/2$, x_1 is not an upper bound, and y_1 is an upper bound.

Now, if $(x_1 + y_1)/2$ is an upper bound for A, let $x_2 = x_1$ and $y_2 = (x_1 + y_1)/2$; otherwise, let $x_2 = (x_1 + y_1)/2$ and $y_2 = y_1$. In either case, the result is $y_2 - x_2 = (b - a)/4$, $x_2 \geq x_1$, x_2 is not an upper bound for A, while $y_2 \leq y_1$, and y_2 is an upper bound for A.

Continuing in this manner, we inductively define an increasing sequence x such that no x_n is an upper bound for A and a decreasing sequence y such that every y_n is an upper bound for A. In addition, $y_n - x_n = (b - a)/2^n$ and y is bounded below by a. Therefore, by hypothesis, y converges to a point $s \in \mathbf{R}$. Furthermore, since $\lim_{n\to\infty} (y_n - x_n) = \lim_{n\to\infty} (b - a)/2^n = 0$, $\lim_{n\to\infty} x_n = s$ by Lemma 7.13.

We claim that s is an upper bound for A. If $z > s$ for some $z \in A$, then $z > y_N$ for some N ⟨*because* $y_n \to s$ ⟩. This contradicts the fact that y_N is an upper bound for A.

Finally, if t is a real number and $t < s$, then $t < x_N$ for some $N \in \mathbf{N}$ by Lemma 7.14. Since x_N is not an upper bound for A, t is not an upper bound. Thus s is a real number which is an upper bound for A and no number less than s is an upper bound; that is, $s = \sup(A)$. Therefore, $\mathbf{R}$ is complete. ■

We have seen in section 7.1 that the field of rationals is not complete. Thus the properties of all three major theorems of this chapter must fail for $\mathbf{Q}$. Since $A = \{x \in \mathbf{Q}: x^2 \leq 2\} = [-\sqrt{2}, \sqrt{2}] \cap \mathbf{Q}$, we say A is closed in $\mathbf{Q}$. A is also bounded but is not compact because $\{(-x, x): x \in A\}$ is an open cover with no finite subcover. The set $B = \{1.4, 1.41, 1.414, 1.4142, 1.41421, \ldots\}$ of (rational) decimal approximations of $\sqrt{2}$ is bounded and infinite but has no accumulation point in $\mathbf{Q}$. Finally, the sequence $x_1 = 1.4$, $x_2 = 1.41$, $x_3 = 1.414$, $x_4 = 1.4142$, $x_5 = 1.41421, \ldots$ is increasing and bounded but fails to converge to any rational number.

Exercises 7.5

1. Prove Lemma 7.13.
2. Prove Lemma 7.14.
3. Give an example of a closed subset A of $\mathbf{Q}$ such that $A \subseteq [7, 8]$ and A is not compact. What open cover of A has no finite subcover?
4. Give a sequence x of distinct rationals such that $\{x_n: n \in \mathbf{N}\} \subseteq [5, 6]$ and x has no accumulation point in $\mathbf{Q}$.

Answers to Selected Exercises

Exercises 1.1

1. (a)

P	$\sim P$	$P \wedge \sim P$
T	F	F
F	T	F

(c)

P	Q	R	$Q \vee R$	$P \wedge (Q \vee R)$
T	T	T	T	T
F	T	T	T	F
T	F	T	T	T
F	F	T	T	F
T	T	F	T	T
F	T	F	T	F
T	F	F	F	F
F	F	F	F	F

(e)

P	Q	$\sim Q$	$P \wedge \sim Q$
T	T	F	F
F	T	F	F
T	F	T	T
F	F	T	F

(g)

P	Q	$\sim Q$	$P \wedge Q$	$(P \wedge Q) \vee \sim Q$
T	T	F	T	T
F	T	F	F	F
T	F	T	F	T
F	F	T	F	T

2. (a) equivalent
   (c) equivalent
   (e) equivalent
   (g) not equivalent
   (i) not equivalent
   (k) equivalent
   (m) equivalent
3. (a) false
   (c) true
   (e) false
   (g) false
   (i) true
   (k) false
4. (a) x is not a positive integer.
   (c) $5 < 3$.

(e) Roses are not red or violets are not blue.
(g) T is green and T is not yellow.

5. (a) If P has the same truth table as Q, then Q has the same truth table as P.
7. (a)

P	Q	$P \textcircled{\vee} Q$
T	T	F
F	T	T
T	F	T
F	F	F

Exercises 1.2

1. (a) Antecedent: squares have three sides.
Consequent: triangles have four sides.
(d) Antecedent: f is differentiable.
Consequent: f is continuous.
(f) Antecedent: f is integrable.
Consequent: f is bounded.
2. (a) Converse: If triangles have four sides, then squares have three sides.
Contrapositive: If triangles do not have four sides, then squares do not have three sides.
(d) Converse: If f is continuous, then f is differentiable.
Contrapositive: If f is not continuous, then f is not differentiable.
(f) Converse: If f is bounded, then f is integrable.
Contrapositive: If f is not bounded, then f is not integrable.
3. (a) true
(c) true
(e) true
4. (a) true
(c) true
5. (b)

P	Q	$\sim P$	$\sim P \Rightarrow Q$	$Q \Leftrightarrow P$	$(\sim P \Rightarrow Q) \vee (Q \Leftrightarrow P)$
T	T	F	T	T	T
F	T	T	T	F	T
T	F	F	T	F	T
F	F	T	F	T	T

(c)

P	Q	$\sim Q$	$Q \Rightarrow P$	$\sim Q \Rightarrow (Q \Leftrightarrow P)$
T	T	F	T	T
F	T	F	F	T
T	F	T	T	T
F	F	T	T	T

6. (f)

P	Q	R	$Q \vee R$	$P \wedge (Q \vee R)$	$P \wedge Q$	$P \wedge R$	$(P \wedge Q) \vee (P \wedge R)$
T	T	T	T	T	T	T	T
F	T	T	T	F	F	F	F
T	F	T	T	T	F	T	T
F	F	T	T	F	F	F	F
T	T	F	T	T	T	F	T
F	T	F	T	F	F	F	F
T	F	F	F	F	F	F	F
F	F	F	F	F	F	F	F

Since the fifth and eighth columns are the same, the propositions $P \wedge (Q \vee R)$ and $(P \wedge Q) \vee (P \wedge R)$ are equivalent.

7. (a) ((f has a relative minimum at x_0) $\wedge$ (f is differentiable at x_0)) $\Rightarrow$ ($f'(x_0) = 0$).
(d) $((x = 1) \vee (x = -1)) \Rightarrow (|x| = 1)$.
(e) (x_0 is a critical point for f) $\Leftrightarrow$ (($f'(x_0) = 0$) $\vee$ ($f'(x_0)$ does not exist)).

8. (b)

P	Q	R	$P \wedge Q$	$P \wedge Q \Rightarrow R$	$\sim R$	$\sim Q$	$P \wedge \sim R$	$P \wedge \sim R \Rightarrow \sim Q$
T	T	T	T	T	F	F	F	T
F	T	T	F	T	F	F	F	T
T	F	T	F	T	F	T	F	T
F	F	T	F	T	F	T	F	T
T	T	F	T	F	T	F	T	F
F	T	F	F	T	T	F	F	T
T	F	F	F	T	T	T	T	T
F	F	F	F	T	T	T	F	T

Since the fifth and ninth columns are the same, the propositions $P \wedge Q \Rightarrow R$ and $P \wedge \sim R \Rightarrow \sim Q$ are equivalent.

9. (a) If x is an even integer, then $x + 1$ is an odd integer.
   (d) not possible.

Exercises 1.3

1. (a) $\sim(\forall x)(x$ is precious $\Rightarrow x$ is beautiful).
   Or, $(\exists x)(x$ is precious and x is not beautiful).
   (c) $(\exists x)(x$ is a positive integer and x is smaller than all other positive integers).
   Or, $(\exists x)(x$ is a positive integer and $(\forall y)(y$ is a positive integer $\Rightarrow x \leq y))$.
   (f) $\sim(\exists x)(\forall y)(x$ loves $y)$ or $(\forall x)(\exists y)(x$ does not love $y)$.
   (h) $(\exists x)(x$ cares about me$)$.
2. (a) $(\forall x)(x$ is precious $\Rightarrow x$ is beautiful). All precious stones are beautiful.
   (c) $(\forall x)(x$ is a positive integer $\Rightarrow ((\exists y)(y$ is a positive integer$) \wedge x > y))$. For every positive integer there is a smaller positive integer.
   Or, $\sim(\exists x)(x$ is a positive integer $\wedge$ $(\forall y)(y$ is a positive integer $\Rightarrow x \leq y.))$ There is no smallest positive integer.
3. *Hint:* To use part (a), note that $\sim(\exists x)(\sim A(x))$ is equivalent to $(\forall x)(\sim\sim A(x))$.
4. (a) true
   (d) false
   (f) false
6. For every backwards E, there exists an upside down A!

Exercises 1.4

1. (a) E. The converse rather than the statement is proved.
   (d) A.

Exercises 1.5

1. (a) E. The false statement referred to is not a denial of the claim.
   (b) C. Uniqueness has not been shown.
   (d) A.

Exercises 2.1

1. (a) $\{x \in \mathbf{N}: x < 6\}$ or $\{x: x \in \mathbf{N}$ and $x < 6\}$.
   (c) $\{x \in \mathbf{R}: 2 \leq x \leq 6\}$ or $\{x: x \in \mathbf{R}$ and $2 \leq x \leq 6\}$.
2. (a) $\{1, 2, 3, 4, 5\}$
   (c) not possible
3. (a) true
   (c) true
   (e) false
   (g) true
   (i) false
4. (a) $\{\{0\}, \{\Delta\}, \{\square\}, \{0, \Delta\}, \{0, \square\}, \{\Delta, \square\}, X, \varnothing\}$
   (c) $\{\{\varnothing\}, \{\{a\}\}, \{\{b\}\}, \{\{a, b\}\}, \{\varnothing, \{a\}\}, \{\varnothing, \{b\}\}, \{\varnothing, \{a, b\}\}, \{\{a\}, \{b\}\}, \{\{a\}, \{a, b\}\}, \{\{b\}, \{a, b\}\}, \{\varnothing, \{a\}, \{b\}\}, \{\varnothing, \{a\}, \{a, b\}\}, \{\varnothing, \{b\}, \{a, b\}\}, \{\{a\}, \{b\}, \{a, b\}\}, X, \varnothing\}$.

5. (a) no proper subsets
   (c) $\{1\}, \{2\}$
6. (a) true
   (c) true
   (e) false
   (g) false
   (i) false
   (k) true
7. (a) $A = \{1, 2\}$, $B = \{1, 2, 4\}$, $C = \{1, 2, 5\}$
   (c) $A = \{1, 2, 3\}$
   (e) not possible
8. (a) true
   (c) true
   (e) true
   (g) true
   (i) true
   (k) true
9. (a) Let x be any object. Let $x \in A$. Then $P(x)$ is true. Since $(\forall x)(P(x) \Rightarrow Q(x))$, $Q(x)$ is true. Thus $x \in B$. Therefore $A \subseteq B$. ■
11. *Hint:* To prove $A = B$, we are given $A \subseteq B$. To prove $B \subseteq A$, use Theorem 2.3.
15. Both $X \in X$ and $X \notin X$ are false because they lead to a contradiction. Thus the collection of all ordinary sets *is not a set*.
16. (a) A. Every statement in the proof is correct even though no reasons are given.
    (c) C. One way to correct the proof would be to insert a second sentence "Let $x \in A$." and a fourth sentence "Then $x \in B$."

Exercises 2.2

1. (a) $\{0, 1, 2, 3, 4, 5, 6, 7, 8, 9\}$
   (c) $\{1, 3, 5, 7, 9\}$
   (e) $\{3, 9\}$
   (g) $\{1, 5, 7\}$
   (i) $\{1, 5, 7\}$
2. (a) $\{0, -2, -4, -6, -8, -10, \ldots\}$
   (c) D
   (e) $N \cup \{0\}$
   (g) D
   (i) $\{0, 2, 4, 6, 8, \ldots\}$
   (k) $\varnothing$
3. (a) $\{\{1\}, \{2\}, \{1, 2\}, \varnothing\}$
   (d) $\{\varnothing, \{1\}, \{3\}\}$
   (f) $\{\{3\}\}$
4. A and B are disjoint.
8. $S \in \mathscr{P}(A \cap B)$ iff $S \subseteq A \cap B$
   iff $S \subseteq A$ and $S \subseteq B$
   iff $S \in \mathscr{P}(A)$ and $S \in \mathscr{P}(B)$
   iff $S \in \mathscr{P}(A) \cap \mathscr{P}(B)$.
10. For all sets A and B, $\varnothing \in \mathscr{P}(A - B)$ but $\varnothing \notin \mathscr{P}(A) - \mathscr{P}(B)$.
12. If C and D are not disjoint, then there is an element x such that $x \in C \cap D$. But then $x \in C$ and $x \in D$. Thus $x \in A$ and $x \in B$, so $x \in A \cap B$. Therefore, A and B are not disjoint.
14. (a) $A = \{1, 2\}$, $B = \{1, 3\}$, $C = \{2, 3, 4\}$.
    (c) $A = \{1, 2\}$, $B = \{1, 3\}$.
    (e) $A = \{1, 2\}$, $B = \{1, 3\}$, $C = \{1\}$.

16. (a) C. The proof that $A \cap B = A$ is incomplete.
    (b) E. The claim is false. The statement "$x \in A$ and $x \in \varnothing$ iff $x \in A$" is false.
    (d) A.
    (f) E. A picture often helps to bring forth ideas around which a correct proof may be built. A proof by picture alone is not sufficient.

Exercises 2.3

1. $\bigcup_{A \in \mathscr{A}} A = \{1, 2, 3, 4, 5, 6, 7, 8\}$; $\bigcap_{A \in \mathscr{A}} A = \{4, 5\}$
3. $\bigcup_{i \in \mathbf{N}} A_i = \mathbf{N}$; $\bigcap_{i \in \mathbf{N}} A_i = \{1\}$
5. $\bigcup_{A \in \mathscr{A}} A = \mathbf{Z}$; $\bigcap_{A \in \mathscr{A}} A = \{10\}$
7. $\bigcup_{n \in \mathbf{N}} A_n = (0, 1)$; $\bigcap_{n \in \mathbf{N}} A_n = \varnothing$
9. $\bigcup_{r \in \mathbf{R}} A_r = [0, \infty)$, $\bigcap_{r \in \mathbf{R}} A_r = \varnothing$
11. The family in exercise 1 is not pairwise disjoint. The family in exercise 2 is pairwise disjoint.
13. (a) If $x \in A_\alpha$ for all $\alpha \in \Delta$, then $x \in A_\beta$.
15. (a)
$$\begin{aligned} x \in B \cap \bigcup_{\alpha \in \Delta} A_\alpha &\text{ iff } x \in B \text{ and } x \in \bigcup_{\alpha \in \Delta} A_\alpha \\ &\text{ iff } x \in B \text{ and } x \in A_\alpha \text{ for some } \alpha \in \Delta \\ &\text{ iff } x \in B \cap A_\alpha \text{ for some } \alpha \in \Delta \\ &\text{ iff } x \in \bigcup_{\alpha \in \Delta} (B \cap A_\alpha). \end{aligned}$$
16. (a)
$$\begin{aligned} \left(\bigcup_{\alpha \in \Delta} A_\alpha\right) \cap \left(\bigcup_{\beta \in \Gamma} B_\beta\right) &= \bigcup_{\beta \in \Gamma} \left(\left(\bigcup_{\alpha \in \Delta} A_\alpha\right) \cap B_\beta\right) \\ &= \bigcup_{\beta \in \Gamma} \left(\bigcup_{\alpha \in \Delta} (A_\alpha \cap B_\beta)\right) \end{aligned}$$
18. (a) Let $x \in \bigcup_{\alpha \in \Gamma} A_\alpha$. Then $x \in A_\alpha$ for some $\alpha \in \Gamma$.
    Since $\Gamma \subseteq \Delta$, $x \in A_\alpha$ for some $\alpha \in \Delta$.
    Therefore, $x \in \bigcup_{\alpha \in \Delta} A_\alpha$. ■
22. Suppose B is a set such that $B \subseteq A_\beta$ for all $\beta \in \Delta$.
    Let $x \in B$. Then, $x \in A_\beta$ for all $\beta \in \Delta$.
    Therefore, $x \in \bigcap_{\beta \in \Delta} A_\beta$.
    Thus $B \subseteq \bigcap_{\beta \in \Delta} A_\beta$.
24. $A_r = (0, r)$, $\mathscr{A} = \{A_r : r \in \mathbf{R}^+\}$
26. (a) A. Note that if we allowed $\Delta = \varnothing$, the claim would be false.
    (b) C. No connection is made between the first and second sentences.

Exercises 2.4

1. Only $\varnothing$ is inductive.
2. (b) true
   (e) false
3. (b) $f_4 = 3$, $f_7 = 13$, $f_{n+3} - f_{n+1} = f_{n+2}$.
6. (a) Let S be a subset of $\mathbf{N}$ such that $1 \in S$ and S is inductive. We wish to show that $S = \mathbf{N}$. Assume that $S \neq \mathbf{N}$ and let $T = \mathbf{N} - S$. By the WOP, the nonempty set T has a least element. This least element is not 1, because $1 \in S$. If the least element is n, then $n \in T$ and $n - 1 \in S$. But by the inductive property of S, $n - 1 \in S$ implies that $n \in S$. This is a contradiction. Therefore, $S = \mathbf{N}$.
11. (b) *Hint:* For each player x, consider the set W_x of all players who win against x.
12. (a) E. Let $S = \{n \in \mathbf{N}$: all horses in every set of n horses have the same color$\}$. Then $1 \in S$. In fact, $S = \{1\}$. The statement $n \in S \Rightarrow n + 1 \in S$ is correct for $n \geq 2$. The only counterexample to $(\forall n)(n \in S \Rightarrow n + 1 \in S)$ occurs when $n = 1$. The "proof" fails to consider that a special argument would be necessary when $n = 1$. In this case there would be no way to remove a different horse from a set of n horses.
    (d) A.

Exercises 3.1

2. (b) $(a, b) \in A \times (B \cap C)$ iff $a \in A$ and $b \in B \cap C$
iff $a \in A$ and $b \in B$ and $b \in C$
iff $a \in A$ and $b \in B$ and $a \in A$ and $b \in C$
iff $(a, b) \in A \times B$ and $(a, b) \in A \times C$
iff $(a, b) \in (A \times B) \cap (A \times C)$. ■

5. (a) domain $\mathbb{R}$, range $\mathbb{R}$
(c) domain $[1, \infty)$, range $[0, \infty)$
(e) domain $\mathbb{R}$, range $\mathbb{R}$.

6. (a) $R_1^{-1} = R_1$.
(c) $R_3^{-1} = \{(x, y) \in \mathbb{R} \times \mathbb{R} : y = \frac{1}{7}(x + 10)\}$.
(e) $R_5^{-1} = \{(x, y) \in \mathbb{R} \times \mathbb{R} : y = \pm\sqrt{(5 - x)/4}\}$.
(g) $R_7^{-1} = \left\{(x, y) \in \mathbb{R} \times \mathbb{R} : y < \dfrac{x + 4}{3}\right\}$.
(i) $R_9^{-1} = \{(x, y) \in \mathbf{P} \times \mathbf{P} : y$ is a child of x, and x is male$\}$.

7. (a) $R_1 \circ R_1 = \{(x, z) : x = z\} = R_1$.
(d) $R_2 \circ R_3 = \{(x, z) \in \mathbb{R} \times \mathbb{R} : z = -35x + 52\}$.
(g) $R_4 \circ R_5 = \{(x, z) \in \mathbb{R} \times \mathbb{R} : z = 16x^4 - 40x^2 + 27\}$.
(j) $R_6 \circ R_6 = \{(x, z) \in \mathbb{R} \times \mathbb{R} : z < x + 2\}$.
(m) $R_3 \circ R_8 = \{(x, z) \in \mathbb{R} \times \mathbb{R} : z = \dfrac{14x}{x - 2} - 10\}$.
(o) $R_9 \circ R_9$ is *not* $\{(x, z) : z$ is a grandfather of $x\}$.

14. (a) E. The statements "$x \in A \times B$" and "$x \in A$ and $x \in B$" are not equivalent.
(b) C. The only error is that $(a, c) \notin B \times D$ implies $a \notin B$ *or* $c \notin D$.
(d) A.

Exercises 3.2

1. (a) not reflexive, not symmetric, transitive
(e) reflexive, not symmetric, transitive
(k) not reflexive, symmetric, not transitive (*Note:* Sibling means "a brother or sister.")

2. (a) $\{(1, 1), (2, 2), (2, 3), (3, 1)\}$.
(d) $\{(1, 1), (2, 2), (3, 3), (1, 3), (2, 3), (3, 1), (3, 2)\}$.

3. (f) This is the graph of the relation $\{(x, y) : y \le x\}$.

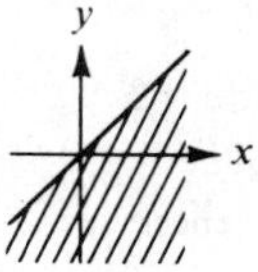

4. (a) The equivalence classes have the form $\{x : x = 2^t n$, where n is odd$\}$, where $t = 0, 1, 2, \ldots$.

5. (a) $0/\equiv_5 = \{\ldots, -15, -10, -5, 0, 5, 10, \ldots\}$
$1/\equiv_5 = \{\ldots, -9, -4, 1, 6, 11, \ldots\}$
$2/\equiv_5 = \{\ldots, -8, -3, 2, 7, 12, \ldots\}$
$3/\equiv_5 = \{\ldots, -7, -2, 3, 8, 13, \ldots\}$
$4/\equiv_5 = \{\ldots, -6, -1, 4, 9, 14, \ldots\}$.

6. (b) Assume R is symmetric. Then $(x, y) \in R$ iff $(y, x) \in R$ iff $(x, y) \in R^{-1}$. Thus $R = R^{-1}$. Now, suppose $R = R^{-1}$. Then $(x, y) \in R$ implies $(x, y) \in R^{-1}$, which implies $(y, x) \in R$. Thus R is symmetric. ■

7. (a) false
(c) false

9. One part of the proof is to show that R is symmetric. Suppose $x\,R\,y$. Then $x\,L\,y$ and $y\,L\,x$, so $y\,L\,x$ and $x\,L\,y$. Therefore, $y\,R\,x$.

10. (b) Assume R is asymmetric. Suppose $(x, y) \in R$ and $(y, x) \in R$. Then, by asymmetry, $(y, x) \notin R$. This is a contradiction. Therefore, $(x, y) \in R$ and $(y, x) \in R$ imply $x = y$.

11. (c) E. The last sentence confuses $R \cap S$ with $R \circ S$. A correct proof requires a more complete second sentence.

Exercises 3.3

2. (a) $\{(1, 1), (1, 2), (2, 1), (2, 2), (3, 3), (3, 4), (3, 5), (4, 3), (4, 4), (4, 5), (5, 3), (5, 4), (5, 5)\}$.
4. No. Let R be the relation $\{(1, 1), (2, 2), (3, 3), (1, 2), (1, 3), (2, 1), (3, 1)\}$ on the set $A = \{1, 2, 3\}$. Then $R(1) = \{1, 2, 3\}$, $R(2) = \{1, 2\}$, and $R(3) = \{1, 3\}$. The set $\mathscr{A} = \{\{1, 2, 3\}, \{1, 2\}, \{1, 3\}\}$ is not a partition of A.
7. (b) Yes. $\{\overline{B_1}, \overline{B_2}\}$ is a partition of A, because $\{\overline{B_1}, \overline{B_2}\} = \{B_2, B_1\}$. If $B_1 = B_2$, then $\overline{B_1} = \overline{B_2} = \emptyset$ so $\{B_1, B_2\}$ is not a partition.
8. (b) C. This is a tough one because the ideas are all there and every statement is true. We give it C because the ideas are not well connected.

Exercises 4.1

1. (a) R_1 is a function. $\text{Dom}(R_1) = \{0, \Delta, \Box, \cap, \cup\}$.
   A possible codomain is $\text{Rng}(R_1) = \{0, \Delta, \Box, \cap, \cup\}$.
   (g) R_7 is a function. $\text{Dom}(R_7) = \mathbf{R}$. Possible codomains are $\mathbf{R}$ and $[0, \infty)$.
2. (a) Domain $= \mathbf{R} - \{1\}$. Range $= \{y \in \mathbf{R} : y \neq 0\}$. A possible codomain is $\mathbf{R}$.
   (d) Domain $= \mathbf{R} - \{\frac{\pi}{2} + k\pi : k \in \mathbf{Z}\}$. A possible codomain is $\mathbf{R}$.
3. (c) 5, -5
4. (a) $\text{Dom}(f) = \mathbf{R} - \{3\}$, $\text{Rng}(f) = \mathbf{R} - \{-1\}$.
8. (a) A.
11. (a) $f(3) = 3/\equiv_6 = \{\ldots, -9, -3, 3, 9, \ldots\}$.

Exercises 4.2

1. (a) $(f \circ g)(x) = 17 - 14x$, $(g \circ f)(x) = -29 - 14x$
   (c) $(f \circ g)(x) = \sin(2x^2 + 1)$, $(g \circ f)(x) = 2\sin^2 x + 1$
2. (a) $\text{Dom}(f \circ g) = \mathbf{R} = \text{Rng}(f \circ g) = \text{Dom}(g \circ f) = \text{Rng}(g \circ f)$.
   (c) $\text{Dom}(f \circ g) = \mathbf{R}$, $\text{Rng}(f \circ g) = [-1, 1]$.
   $\text{Dom}(g \circ f) = \mathbf{R}$, $\text{Rng}(g \circ f) = [1, 3]$.
3. (a) $f^{-1}(x) = \dfrac{x - 2}{5}$
   (c) $f^{-1}(x) = \dfrac{1 - 2x}{x - 1}$
   (e) $f^{-1}(x) = -3 + \ln x$
7. (a) $\{(x, y) \in \mathbf{R} \times \mathbf{R} : y = 0 \text{ if } x < 0 \text{ and } y = x^2 \text{ if } x \geq 0\}$
   $\{(x, y) \in \mathbf{R} \times \mathbf{R} : y = x^2\}$.
11. (a) $h \cup g$ is a function.

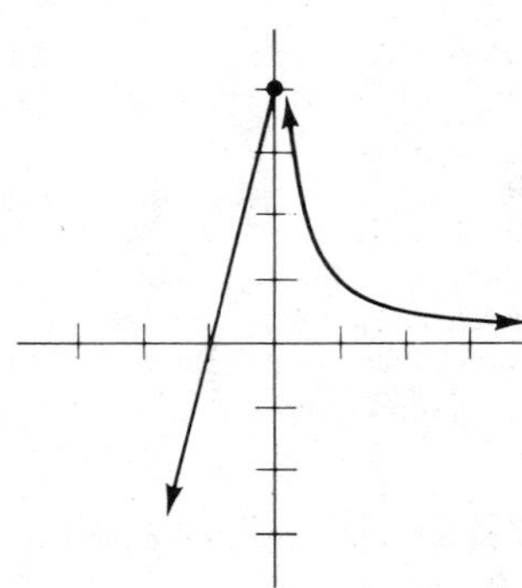

12. *Hint:* Write $A \cup C$ as $A \cup (C - E)$. Then show $h \cup g\,|_{C-E}$ and use Theorem 4.6.
13. (a) A.

Exercises 4.3

1. (a) Onto $\mathbf{R}$. Let $w \in \mathbf{R}$. Then for $x = 2(w - 6)$, we have $f(x) = \frac{1}{2}(2(w - 6)) + 6 = w$. Thus $w \in \text{Rng}(f)$. Therefore, f maps onto $\mathbf{R}$. ■
   (c) Not onto $\mathbf{N} \times \mathbf{N}$. Since $(5, 8) \in \mathbf{N} \times \mathbf{N}$ and $(5, 8) \notin \text{Rng}(f)$, f does not map onto $\mathbf{N} \times \mathbf{N}$.
2. (a) One-to-one. Suppose $f(x) = f(y)$. Then $\frac{1}{2}x + 6 = \frac{1}{2}y + 6$. Then $\frac{1}{2}x = \frac{1}{2}y$, so $x = y$.
   (g) Not one-to-one, because $\sin(\frac{\pi}{6}) = \sin(\frac{13\pi}{6}) = \frac{1}{2}$.
3. (a) $B = \{0, 3\}$, $f = \{(1, 0), (2, 3), (3, 0), (4, 0)\}$.
8. (a) Let $f: \mathbf{R} \to \mathbf{R}$ be given by $f(x) = 2x$ and $g: \mathbf{R} \to \mathbf{R}$ be given by $g(x) = x^2$. Then f maps onto $\mathbf{R}$ but $g \circ f$ is not onto $\mathbf{R}$.
   (e) Let $A = \{a, b, c\}$, $B = \{1, 2, 3\}$, $C = \{x, y, z\}$, $f = \{(a, 2), (b, 2), (c, 3)\}$, and $g = \{(1, x), (2, y), (3, z)\}$. Then g is one-to-one, but $g \circ f = \{(a, y), (b, y), (c, z)\}$ is not.
10. (a) Let $A = B = \mathbf{R}$ and $S = \{(x, y) \in \mathbf{R} \times \mathbf{R}: x^2 + y^2 = 25\}$. Then $(3, 4) \in S$ and $(3, -4) \in S$, but $f_1(3, 4) = 3 = f_1(3, -4)$.
11. (a) Yes.
    (c) No. The projection f_2 is one-to-one iff S is one-to-one.
12. (c) A.
    (d) E. The "proof" does not show that every $w \in (-\frac{\pi}{2}, \frac{\pi}{2})$ is in the range.

Exercises 4.4

1. (a) $(\varnothing, \varnothing)$, $(\{1\}, \{4\})$, $(\{2\}, \{4\})$, $(\{3\}, \{5\})$, $(\{1, 2\}, \{4\})$, $(\{1, 3\}, \{4, 5\})$, $(\{2, 3\}, \{4, 5\})$, $(A, \{4, 5\})$
2. (a) $[2, 10]$
   (c) $\{0\}$
4. (a) $\{(1, 1)\}$
5. (b) $\{p, s, t\}$
   (d) $\{1, 2, 3, 5\}$
6. (c) $\left[2 - \sqrt{3}, \dfrac{3 - \sqrt{5}}{2}\right) \cup \left(\dfrac{3 + \sqrt{5}}{2}, 2 + \sqrt{3}\right]$
7. (a) $[9, 25]$
12. (b) Let $t \in f(X) - f(Y)$. Then $t \in f(X)$, so there exists $x \in X$ such that $f(x) = t$. We note $x \notin Y$ since $t = f(x) \notin f(Y)$. Thus $x \in X - Y$ and, therefore, $t = f(x) \in f(X - Y)$.
13. The converse is true. To prove f is one-to-one, let $x \neq y$ in A. Then $\{x\} \cap \{y\} = \varnothing$ and thus $f(\{x\} \cap \{y\}) = \varnothing$. By hypothesis, $f(\{x\} \cap \{y\}) = f(\{x\}) \cap f(\{y\}) = \{f(x)\} \cap \{f(y)\}$. Thus $f(x) \neq f(y)$.
16. (a) If f is one-to-one, then the induced function is one-to-one.
18. (a) E. The claim is not true. We cannot conclude $x \in X$ from $f(x) \in f(X)$.

Exercises 5.1

3. *Hint:* Consider $f: A \to A \times \{x\}$ given by $f(a) = (a, x)$ for each $a \in A$.
4. (a) finite
   (c) finite
   (e) infinite
6. (a) Suppose A is finite. Since $A \cap B$ is a subset of A, $A \cap B$ is finite.
7. (a) *Hint:* Consider $f: \mathbf{N}_k \times \mathbf{N}_m \to \mathbf{N}_{km}$ given by $f(a, b) = (a - 1)m + b$. Use the Division Algorithm to show f is a one-to-one correspondence.
9. (c) not possible
10. *Hint:* Write $A \cup B = (A - B) \cup B$ and $A = (A - B) \cup (A \cap B)$ and apply Theorem 5.6.
11. *Hint:* Use the Principle of Mathematical Induction on the number n. Begin with $n = 2$. In the inductive step, assume there is a one-to-one function $f: \mathbf{N}_{n+1} \to \mathbf{N}_r$. Consider $f|_{\mathbf{N}_n}$ and the two cases $r < n$ and $r = n$.
12. *Hint:* Suppose $f: A \xrightarrow[\text{onto}]{1-1} \mathbf{N}_k$. If $k \notin A$, use exercise 11 to get a contradiction. If $k \in A$, choose $x \in \mathbf{N}_k - A$. Use exercise 11 and the function $(f - \{(k, f(k))\} \cup \{(x, f(k))\}$ to get a contradiction.
15. *Hint:* Let S have cardinal number n and cardinal number m. Then $S \approx \mathbf{N}_n$ and $S \approx \mathbf{N}_m$. Thus $\mathbf{N}_n \approx \mathbf{N}_m$. Now use exercise 11.
16. Use induction on the number of elements in the domain.
17. (b) C. In case 2, $\mathbf{N}_k \cup \{x\} \approx \mathbf{N}_{k+1}$, but $\mathbf{N}_k \cup \{x\} \not\approx \mathbf{N}_k \cup \mathbf{N}_1 \not\approx \mathbf{N}_{k+1}$.

Exercises 5.2

1. (a) Let $f\colon \mathbf{N} \to D^+$ be given by $f(n) = 2n - 1$ for each $n \in \mathbf{N}$. We show that f is one-to-one and maps onto D^+. First, to show f is one-to-one, suppose $f(x) = f(y)$. Thus $2x - 1 = 2y - 1$, which implies $x = y$. Also, f maps onto D^+ since if d is an odd positive integer, then d has the form $d = 2r - 1$ for some $r \in \mathbf{N}$. But then $f(r) = d$.
   (e) *Hint:* Consider $f(x) = -(x + 12)$ with domain $\mathbf{N}$.
4. (b) Let $f\colon (0, 1) \to (4, 6)$ be given by $f(x) = 2x + 4$ for each $x \in (0, 1)$. Show f maps onto $(4, 6)$: Let $t \in (4, 6)$. Then $4 < t < 6$. Thus $0 < t - 4 < 2$, and so $0 < \dfrac{t-4}{2} < 1$. Let $r = \dfrac{t-4}{2}$. Then $r \in (0, 1)$ and $f(r) = t$. Now, show f is one-to-one. Let $f(x) = f(y)$. Then $2x + 4 = 2y + 4$, which implies $x = y$.
5. (c) Let $f(x) = 1/x$.
8. (a) c
   (c) $\aleph_0$
   (e) c
9. (b) Let m be the largest number in S. Then $\mathbf{N} - S = (\mathbf{N} - \mathbf{N}_m) \cup (\mathbf{N}_m - S)$. By part (a) $\mathbf{N} - \mathbf{N}_m$ is denumerable. Since $\mathbf{N}_m - S$ is a subset of $\mathbf{N}_m$, $\mathbf{N}_m - S$ is finite. Therefore, by Theorem 5.14, $\mathbf{N} - S$ is denumerable.
10. *Hint:* Mimic the proof that $(0, 1)$ is uncountable.
12. *Hint:* Use induction on n and Theorem 5.15.
14. (a) E. The main idea of the "proof" is that infinite subsets of $\mathbf{N}$ are denumerable. No justification for this is given.
   (c) E. The claim is false. Also "A and B are finite" is not a denial of "A and B are infinite."
   (d) E. Writing an infinite set A as $\{x_1, x_2, \ldots\}$ is the same as assuming A is denumerable.

Exercises 5.3

1. We must show that $n \le \aleph_0$ and $n \ne \aleph_0$. The inclusion map $i\colon \mathbf{N}_n \to \mathbf{N}$ is one-to-one; hence, $\overline{\overline{\mathbf{N}_n}} \le \overline{\overline{\mathbf{N}}}$. If $n = \aleph_0$, then $\overline{\overline{\mathbf{N}_n}} = \overline{\overline{\mathbf{N}}}$, and thus $\mathbf{N}_n \approx \mathbf{N}$. But $\mathbf{N}$ is not finite. Therefore, $n \ne \aleph_0$. Thus $n < \aleph_0$.
5. (a) $\overline{\overline{\varnothing}} < \overline{\overline{\{0\}}} < \overline{\overline{\{0, 1\}}} < \overline{\overline{\mathbf{Q}}} < \overline{\overline{(0, 1)}} = \overline{\overline{[0, 1]}} = \overline{\overline{\mathbf{R} - \mathbf{Q}}} = \overline{\overline{\mathbf{R}}} < \overline{\overline{\mathscr{P}(\mathbf{R})}} < \overline{\overline{\mathscr{P}(\mathscr{P}(\mathbf{R}))}}$.
7. *Hint:* There is a proof by contradiction using Cantor's Theorem.
10. (b) not possible
13. (b) E. We have not defined or discussed properties of operations such as addition for cardinal numbers. Thus $\overline{\overline{C}} = \overline{\overline{B}} + (\overline{\overline{C - B}})$ cannot be used unless C and B are finite sets.
   (d) A.

Exercises 5.4

2. Let $B \subseteq A$ with B infinite and A denumerable. Since $B \subseteq A$, $\overline{\overline{B}} \le \overline{\overline{A}}$. Since A is denumerable, $\overline{\overline{A}} = \overline{\overline{\mathbf{N}}}$. Since B is infinite, B has a denumerable subset D by Theorem 5.23. Thus $\overline{\overline{A}} = \overline{\overline{\mathbf{N}}} = \overline{\overline{D}} \le \overline{\overline{B}}$. By the Cantor-Schröder-Bernstein Theorem, $\overline{\overline{B}} = \overline{\overline{A}}$. Thus $B \approx A$.
5. (a) A.
   (c) A.

Exercises 5.5

1. (a) Let $A_n = \{n\}$ for each $n \in \mathbf{Z}$. Each A_n is finite and $\bigcup_{n \in \mathbf{Z}} A_n = \mathbf{Z}$, which is denumerable.
6. (a) *Hint:* Write $\mathbf{N} \times \mathbf{N} = \bigcup_{n \in \mathbf{N}} A_n$, where $A_n = \{(m, n)\colon m \in \mathbf{N}\}$ for each $n \in \mathbf{N}$. Show A_n is denumerable for all $n \in \mathbf{N}$ and then use Theorem 5.28.

7. *Hint:* Since $\mathbf{N} \subseteq \mathbf{Q}$, $\overline{\overline{\mathbf{N}}} \le \overline{\overline{\mathbf{Q}}}$. On the other hand, the function $f\colon \mathbf{Q} \to \mathbf{Z} \times \mathbf{Z}$, given by $f(p/q) = (p, q)$, can be used to show $\overline{\overline{\mathbf{Q}}} \le \overline{\overline{\mathbf{N}}}$.
8. (b) *Hint:* Let $T = \{\{m, n\}\colon m, n \in \mathbf{N}$ with $m \neq n\}$. Use $g\colon \mathbf{N} \to T$, defined by $g(x) = \{x, x+1\}$ for all $x \in \mathbf{N}$, to show $\overline{\overline{\mathbf{N}}} \le \overline{\overline{T}}$. Then use $f\colon T \to \mathbf{N} \times \mathbf{N}$, given by $f(\{m, n\}) = (\min\{m, n\}, \max\{m, n\})$ for all $\{m, n\} \in T$, to show $\overline{\overline{T}} \le \overline{\overline{\mathbf{N} \times \mathbf{N}}}$.
9. *Hint:* For each $n \in \mathbf{N}$, let T_n be the set of all sequences where all but n terms are 0. Show that T_n is denumerable (using exercise 8 (d)) and that $S' = \bigcup_{n \in \mathbf{N}} T_n$.
11. (a) C. The proof is valid only when $f(1) = x$. In the case when $f(1) \neq x$, we must first redefine f: let t be the unique element of $\mathbf{N}$ such that $f(t) = x$ and define $\hat{f} = (f - \{(1, f(1)), (t, x)\}) \cup \{(1, x), (t, f(1))\}$. Now let $g(n) = \hat{f}(n+1)$ for all $n \in \mathbf{N}$.

Exercises 6.1

1. (a) yes
   (e) no
2. (a) not commutative
   (e) not an operation
3. (a) not associative
   (e) not an operation
8. *Hint:* Compute $e \circ f$.
9. (a) *Hint:* Compute $x \circ (a \circ y)$ and $(x \circ a) \circ y$.
15. $I(f+g) = \int_a^b (f+g)(x)dx = \int_a^b f(x)dx + \int_a^b g(x)dx = I(f) + I(g)$.
21. (a) Let $C, D \in \mathscr{P}(A)$. Then $f(C \cup D) = f(C) \cup f(D)$ by Theorem 4.16 (a). Therefore, f is an OP mapping.
22. (b) E. The claim is false. One may pre-multiply (multiply on the left) or post-multiply both sides of an equation by equal quantities. Multiplying one side on the left and the other on the right is not allowed.

Exercises 6.2

1. (a)

$\cdot$	1	-1	i	$-i$
1	1	-1	i	$-i$
-1	-1	1	$-i$	i
i	i	$-i$	-1	1
$-i$	$-i$	i	1	-1

We see from the table that the set is closed under $\cdot$, 1 is the identity, and each element has an inverse. Also, $\cdot$ is associative.

   (d) *Hint:* $\varnothing$ is the identity.

2.

	e	u	v	w
e	e	u	v	w
u	u	v	w	e
v	v	w	e	u
w	w	e	u	v

4. (a) The group is abelian.
7. *Hint:* For $a, b \in G$, compute a^2b^2 and $(ab)^2$ two ways.
8. *Hint:* In order to have both cancellation properties, every element must occur in every row and in every column of the table.
11. *Hint:* Let $a, b \in G$. For an element x such that $a * x = b$, try $a^{-1} * b$.
13. (b) E. A minor criticism is that no special case is needed for e. The fatal flaw is the use of the undefined division notation.

Exercises 6.3

3. (i) Suppose $a \equiv_m b$ and $c \equiv_m d$. Then m divides $a - b$ and $c - d$. Thus there exist integers k and ℓ such that $a - b = km$ and $c - d = \ell m$. Hence $(a + c) - (b + d) = (a - b) + (c - d) = km + \ell m = (k + \ell)m$. Then $k + \ell$ is an integer so m divides $(a + c) - (b + d)$. Therefore, $a + c \equiv_m b + d$.
6. (b) (143256)
7. (b) $\alpha^2 = (312)$ and $\alpha^3 = (123)$, the identity. Therefore, $\alpha^{-1} = \alpha^2$; $\alpha^4 = \alpha^3\alpha = \alpha$; $\alpha^{50} = \alpha^{48}\alpha^2 = (\alpha^3)^{12}\alpha^2 = \alpha^2$; $\alpha^{51} = \alpha^{50}\alpha = \alpha^2\alpha = \alpha^3 = (123)$.
10. 10, 12, $2n$.
12. (a) The zero divisors are 0, 2, 3, 4, 6, 8, 9, 10.
13. Since $(p - 1)(p - 1) = p^2 - 2p + 1 = p(p - 2) + 1$, $(p - 1)^2 \equiv_p 1$. Therefore $(p - 1)(p - 1) = 1$ in $\mathbf{Z}_p$, and hence $(p - 1)^{-1} = p - 1$.
14. (a) $x = 0, 4, 8, 12, 16$.
15. (b) E. There is no justification that $xy \neq 0$.

Exercises 6.4

1. Each group has four subgroups.
9. N_a is not empty, because $ea = a = ae$, so $e \in N_a$. Let $x, y \in N_a$. Then $xa = ax$ and $ya = ay$. Multiplying both sides of the last equation by y^{-1}, we have $y^{-1}(ya)y^{-1} = y^{-1}(ay)y^{-1}$. Thus $(y^{-1}y)(ay^{-1}) = (y^{-1}a)(yy^{-1})$, or $ay^{-1} = y^{-1}a$. Therefore, $(xy^{-1})a = x(y^{-1}a) = x(ay^{-1}) = (xa)y^{-1} = (ax)y^{-1} = a(xy^{-1})$. This shows $xy^{-1} \in N_a$. Therefore, N_a is a subgroup of G, by Theorem 6.11.
15. The identity $e \in H$ because H is a group, and thus $e^{-1}ee \in K$. Thus K is not empty. Suppose $b, c \in K$. Then $b = a^{-1}h_1a$ and $c = a^{-1}h_2a$ for some $h_1, h_2 \in H$. Thus $bc^{-1} = (a^{-1}h_1a)(a^{-1}h_2a)^{-1} = (a^{-1}h_1a)(a^{-1}h_2^{-1}a) = a^{-1}h_1(aa^{-1})h_2^{-1}a = a^{-1}h_1h_2^{-1}a$. But H is a group, so $h_1h_2^{-1} \in H$. Thus $bc^{-1} \in H$. Therefore, K is a subgroup of G.
19. (a) 1, 3, 5, or 15
22. (c) a^{10}, a^{20}
25. *Hint:* Use exercise 21 (a).

Exercises 6.5

1. $H = \{(213), (123)\}$. The left cosets of H are $eH = H$, $(132)H = \{(132), (312)\}$, and $(321)H = \{(321), (213)\}$.
4. Let G be a group of order 4. By Corollary 6.17, the order of an element of G must be 1, 2, or 4. The only element of order 1 is the identity. If G has an element of order 4, then G is cyclic. Otherwise, every element (except the identity) has order 2.
9. *Hint:* Let $G = [a]$, and suppose $n = mk$. Consider the element a^k.
10. (a) E. From $aH = bH$ we can conclude only that for some h_1 and $h_2 \in H$, $ah_1 = bh_2$.
    (c) A.

Exercises 6.6

1. (i) Suppose H is normal. Let $a \in G$ and show $a^{-1}Ha \subseteq H$. Let $t \in a^{-1}Ha$. Then $t = a^{-1}ha$ for some $h \in H$. Thus $at = ha$. Since $ha \in Ha$, $at \in Ha$. By normality $Ha = aH$. Therefore, $at \in aH$, so there exists $k \in H$ such that $at = ak$. By cancellation, $t = k$. Thus $t \in H$.
   (ii) We let $x \in G$ and show that $xH = Hx$. Let $y \in xH$. Then $y = xh$ for some $h \in H$. But then $yx^{-1} = xhx^{-1}$. Since $xhx^{-1} \in xHx^{-1} \subseteq H$, $yx^{-1} \in H$. Therefore, $yx^{-1} = k$ for some $k \in H$. Thus $y = kx$, which proves $y \in Hx$. Therefore, $xH \subseteq Hx$. A proof that $Hx \subseteq xH$ is similar.
2. *Hint:* Use exercise 1.
5. (b) *Hint:* If $G = [a]$, the group generated by $a \in G$, consider $[aH]$, the group generated by the element aH in G/H. As a lemma prove that if $x \in G$, then $(xH)^n = x^nH$ for all $n \in \mathbf{Z}$.
8. *Hint:* See exercise 7.
12. (b) C. The proof is correct but a verification that $x(H \cap K) = xH \cap xK$ is not provided. Such a verification might not be required in an advanced class.

Exercises 6.7

5. *Hint:* See exercise 25 of section 6.4.

Exercises 7.1

1. (a) supremum: 1; infimum: 0
   (c) supremum does not exist; infimum: 0
   (e) supremum: 1; infimum: $\frac{1}{3}$
   (g) supremum: 5; infimum: -1
4. (a) Let x and y be least upper bounds for A. Then x and y are upper bounds for A. Since y is an upper bound and x is a least upper bound $x \leq y$. Since x is an upper bound and y is a least upper bound $y \leq x$. Thus $x = y$.
9. (a) Let $s = \sup(A)$ and $t = \inf(B)$, where $B = \{u: u \text{ is an upper bound for } A\}$. We must show $s = t$.
   (1) To show $t \leq s$ we note that since $s = \sup(A)$, s is an upper bound for A. Thus $s \in B$. Therefore, $t \leq s$.
   (2) To show $s \leq t$ we will show t is an upper bound for A. If t is not an upper bound for A, then there exists $a \in A$ with $a > t$. Let $\epsilon = (a - t)/2$. Since $t = \inf(B)$ and $t < t + \epsilon$, there exists $u \in B$ such that $u < t + \epsilon$. But $t + \epsilon < a$. Therefore, $u < a$, a contradiction, since $u \in B$ and $a \in A$.
12. (a) Let $m = \max\{\sup(A), \sup(B)\}$.
   (1) Since $A \subseteq A \cup B$, $\sup(A) \leq \sup(A \cup B)$. Also, $B \subseteq A \cup B$ implies $\sup(B) \leq \sup(A \cup B)$. Thus $m = \max\{\sup(A), \sup(B)\} \leq \sup(A \cup B)$.
   (2) It suffices to show m is an upper bound for $A \cup B$. Let $x \in A \cup B$. If $x \in A$, then $x \leq \sup(A) \leq m$. If $x \in B$, then $x \leq \sup(B) \leq m$. Thus m is an upper bound for $A \cup B$. Hence $\sup(A \cup B) \leq m$.
13. (a) E. The claim is true but $y = i + \epsilon/2$ need not be in A.

Exercises 7.2

1. (b) open
   (e) open
   (i) closed
2. *Hint:* Use De Morgan's Laws.
5. If $A = \{x_1, x_2, \ldots, x_n\}$ with $x_1 < x_2 < x_3 < \cdots < x_n$, then $\tilde{A} = (-\infty, x_1) \cup (x_1, x_2) \cup \cdots \cup (x_{n-1}, x_n) \cup (x_n, \infty)$. Thus $\tilde{A}$ is open since it is a union of open sets. Therefore, A is closed.
10. (a) *Hint:* First show that an open cover of $A \cup B$ is an open cover of A and an open cover of B. A different proof uses the Heine-Borel Theorem.
13. (a) not compact (not bounded)
   (d) not compact (neither closed nor bounded)
   (g) not compact (not closed)
14. *Hints:* One proof uses directly the definition of compactness; the other uses the Heine-Borel Theorem.
18. (b) C. With the addition of O^* to the cover $\{O_\alpha: \alpha \in \Delta\}$ we are assured that there is a finite subcover of $\{O^*\} \cup \{O_\alpha: \alpha \in \Delta\}$, but not necessarily a subcover of $\{O_\alpha: \alpha \in \Delta\}$. Since $O^* = A - B$ is useless in a cover of B, it can be deleted from the subcover after it is used.

Exercises 7.3

1. (a) $\{\frac{1}{2}\}$
   (c) $\varnothing$
   (g) $\{0, 2\}$
3. *Hint:* First show for all $a \in A$, $z > a$. Then show z is an accumulation point of A by using Theorem 7.1.
6. *Hint:* Show $(A \cap B)' \subseteq A' \cap B'$ by using exercise 4 (a).
8. (c) By parts (a) and (b), $c(A)$ is closed and contains A. Let $A \subseteq B$, and let B be closed. Then, by exercise 7, $A' \subseteq B$. Thus, $c(A) = A \cup A' \subseteq B$.
9. (a) has no accumulation points
   (c) has accumulation points
12. $\{0, 1\}$

14. (a) E. *Hint:* See page 28 on the misuse of quantifiers.
    (c) E. $(\overline{B})'$ need not be a subset of $(\overline{B'})$.

Exercises 7.4

1. (a) $x_n \to 0$; for $\epsilon > 0$, use $N > 1/\epsilon$
   (c) x diverges; use $\epsilon = 1$
   (f) $x_n \to 0$; for $\epsilon > 0$, use $N > (2\epsilon)^{-2}$ and $\sqrt{n+1} - \sqrt{n} = (\sqrt{n+1} - \sqrt{n})\left(\dfrac{\sqrt{n+1} + \sqrt{n}}{\sqrt{n+1} + \sqrt{n}}\right)$.
4. (a) Let $\epsilon > 0$. Then $\epsilon/2 > 0$. Since $x_n \to L$, there exists $N_1 \in \mathbf{N}$ such that if $n > N_1$, then $|x_n - L| < \epsilon/2$. Likewise, there exists $N_2 \in \mathbf{N}$ such that $n > N_2$ implies $|y_n - M| < \epsilon/2$. Let $N_3 = \max\{N_1, N_2\}$, and assume $n > N_3$. Then $|(x_n + y_n) - (L + M)| = |(x_n - L) + (y_n - M)| \leq |x_n - L| + |y_n - M| < \epsilon/2 + \epsilon/2 = \epsilon$. Therefore, $x_n + y_n \to L + M$.
5. *Hint:* If y is bounded by a positive number B, use the definition of $x_n \to 0$ with ϵ/B.
6. (a) *Hint:* $||x_n| - |L|| \leq |x_n - L|$.
9. (a) *Hint:* Since $x_n \to L$, $|x_n| \to |L|$. Now apply the definition of $|x_n| \to |L|$ with $\epsilon = |L|/2$.
10. (b) A. The proof uses exercise 4(b).

Index

Abelian group, 120
Accumulation point, 154
Algebraic structure (*see* Algebraic system)
Algebraic system, 114
Antecedent, 5
Appel, Kenneth, 23
Associative operation, 115
Axiom of Choice, 108–109

Bernstein, Felix, 105
Biconditional sentence, 7
  translation of, 10
Bijection (*see* One-to-one corespondence)
Binary operation:
  definition, 114
  set closed under, 115
Bolzano, Bernard, 154
Bolzano-Weierstrass Theorem, 154–155, 160–161
Borel, Emile, 151
Bounded-Monotone Sequence Theorem, 158–159, 160–161
Bounded sequence, 156
Bounded set, 141

Cantor, Georg, 100, 103, 104, 105
Cantor-Schröder-Bernstein Theorem, 105–106
Cantor's Theorem, 104–105
Cardinal number, 93, 97, 99, 103
Cartesian product, 54
Cayley, Arthur, 139
Cayley table (*see* Operation table)
Cayley's Theorem, 139
Center of a group, 132
Closed:
  interval, 31
  ray, 31
  set, 148
Closure of a set, 155
Codomain, 71
Commutative operation, 115
Compact, 150
Comparability Theorem, 108
Complement, 38
Complete field, 146, 160–161
Composition:
  of a function, 75–76
  of a relation, 58
Conditional sentence: 5–6
  contrapositive of, 6
  converse of, 6
  translation of, 9–10
Congruence modulo *m*, 64, 126–127
Conjunction, 1–2
Consequent, 5
Contrapositive, 6
Converse, 6
Coset, 133–137
Countable set, 97, 111–113
Counterexample, 25
Cover of a set, 150
Cross product, 54
Cyclic group, 130

Denial, 4
Denumerable set, 97, 100–101, 110
Derived set, 154
De Morgan's Laws, 16, 39, 44
Difference of sets, 36
Disjoint sets, 36
Disjunction, 1–2
Division Algorithm:
  for Integers, 53
  for natural numbers, 51
Domain, 56

Empty set, 31
Equality of sets, 34
Equivalence:
  class, 63
  relation, 63
Equivalent:
  propositions, 3
  quantified sentences, 13
  sets, 93
Euclid, 21, 23

Euler, Leonard, 24
Exclusive or, 5
Existential quantifier, 12–14

Family of sets, 41
Field:
  complete, 146
  definition, 143
  ordered, 144
Finite set, 93
Function:
  canonical, 72, 130
  characteristic, 72
  constant, 72
  definition, 70
  extension of, 77
  identity, 73
  inclusion, 73
  one-to-one, 83–85
  onto, 81, 83–85
  restriction of, 77, 85
  union of, 78, 85
Fundamental Theorem of Group Homomorphisms, 141

Galois, Evariste, 120
Generator, 130
Graph, 56
Greatest lower bound (*see* Infimum)
Group:
  abelian, 120
  cyclic, 130
  definition, 120
  isomorphic, 139–141
  octic, 125
  of permutations, 124
  symmetric, 124

Haken, Wolfgang, 23
Heine, Edward, 151
Heine-Borel Theorem, 151–152, 160–161
Homomorphic image, 122
Homomorphism, 122
Hypothesis of induction, 49

Identity:
  element, 115
  relation, 57
Image:
  of an element, 71
  of a set, 87
Inclusive or (*see* Disjunction)
Indexed family, 43
Index:
  of a group, 134
  of a set, 43
Inductive set, 47
Infimum, 145
Infinite set, 93
Injection (*see* One-to-one function)
Interior point, 148

Intersection:
  of two sets, 36
  over a family, 41, 43
Interval notation, 31
Inverse:
  element, 115
  of a function, 75, 83–85
  of a relation, 59
Inverse image of a set, 87
Isomorphism, 139–141

Kernel, 129

Lagrange's Theorem, 134
Least upper bound (*see* Supremum)
Limit of a sequence, 157
Lower bound, 141

Mapping (*see* Function)
Monotone sequence, 158
Multiple of an element, 122

Negation, 1–2
Negative of an element, 122
Neighborhood, 147
Normalizer, 132
Normal subgroup, 136

Octic group, 125
One-to-one correspondence, 84, 93
One-to-one function, 83–85
Onto function, 81, 83–85
Open:
  interval, 31
  ray, 31
  sentence, 11
  set, 148
Operation preserving mapping, 116
Operation table, 115
Order:
  of an algebraic system, 115
  of an element, 130
Ordered n-tuple, 54
Ordered pair, 54

Pairwise disjoint, 44
Partition, 66
Permutation group, 124, 139
Permutation, 124
Power set, 32
Pre-image under a function, 71
Principle of Complete Induction, 50
Principle of Mathematical Induction, 47
Proof:
  of biconditionals, 22–23
  by contradiction, 21–22, 26
  by contrapositive, 20–21, 24–25
  by exhaustion, 23
  constructive, 24
  definition, 16
  direct, 18–20, 26

Proposition:
- biconditional, 7
- conditional, 5–6
- definition, 1
- denial, 4
- equivalent, 3

Quantifier:
- existential, 12–14
- unique existential, 15
- universal, 12–14

Quotient group, 135–141

Range, 56
Reflexive relation, 62
Relation:
- definition, 55
- equivalence, 63
- reflexive, 62
- symmetric, 62
- transitive, 62

Schröder, Ernest, 105
Sequence:
- bounded, 156
- convergent, 157
- decreasing, 158
- definition, 156
- divergent, 157
- increasing, 158
- limit of, 157
- monotone, 158

Set notation, 30
Subcover, 150
Subgroup:
- definition, 128
- identity, 128
- normal, 136
- proper, 128

Subset:
- definition, 31
- improper, 32
- proper, 32

Supremum, 145
Surjection (*see* Onto function)
Symmetric group, 124
Symmetric relation, 62
Symmetry, 124

Tautology, 16
Theorem, 16
Transitive relation, 62
Truth set, 11

Uncountable set, 97–99
Undecidable sentence, 23
Union:
- of two sets, 36
- over a family, 41, 43

Universal quantifier, 12–14
Universe of discourse, 11
Upper bound, 141

Weierstrass, Karl, 154
Well Ordering Principle, 51

List of Symbols

$P \wedge Q$, $P \vee Q$, $\sim P$ 1

$P \Rightarrow Q$ 5

$P \Leftrightarrow Q$ 7

$(\forall x)P(x)$, $(\exists x)P(x)$ 12

$(\exists! x)P(x)$ 15

$\mathbf{N}$, $\mathbf{Z}$, $\mathbf{Q}$, $\mathbf{R}$ 31

$\varnothing$ 31

$x \in A$ 31

$A \subseteq B$ 31

$[a, b]$, (a, b), (a, ∞), $(-\infty, a)$ 31

$\mathscr{P}(A)$ 32

$A = B$ 34

$A \cup B$, $A \cap B$, $A - B$ 36

$\tilde{A}$ 38

$\bigcup_{A \in \mathscr{A}} A$, $\bigcap_{A \in \mathscr{A}} A$ 41

$\bigcup_{\alpha \in \Delta} A_\alpha$, $\bigcap_{\alpha \in \Delta} A_\alpha$ 43

PMI 47

PCI 50

WOP 51

$A \times B$ 54

$a R b$, $a \not R b$ 56

$\mathrm{Dom}(R)$, $\mathrm{Rng}(R)$ 56

I_A 57, 73

R^{-1} 58

$S \circ R$ 58

x/R, A/R 63

$\equiv_m$ 64

$\mathbf{Z}_m$ 64, 126

$f: A \rightarrow B$ 70

χ_A 72

$f|_D$ 77

$f: A \xrightarrow{\text{onto}} B$ 81

$f: A \xrightarrow{1-1} B$ 83

$f(X)$, $f^{-1}(Y)$ 87

$A \approx B$, $A \not\approx B$ 93

$\mathbf{N}_k$ 93

$\bar{\bar{A}}$ 93

$\aleph_0$ 97

c 99

$\mathbf{Q}^+$ 100

$\bar{\bar{A}} = \bar{\bar{B}}$, $\bar{\bar{A}} \leq \bar{\bar{B}}$, $\bar{\bar{A}} < \bar{\bar{B}}$ 104

$(A, *)$ 115

OP 116

S_n 124

$\mathscr{O}$ 125

$+_m$, $\cdot_m$ 126

$\ker(f)$ 129

$[a]$ 130

xH, Hx 133

G/H 135

$\sup(A)$, $\inf(A)$ 145

$\mathscr{N}(a, \delta)$ 147

x_n 156

$\lim_{n \to \infty} x_n = L$, $x_n \rightarrow L$ 157